The Disordered Police State

The Disordered Police State

GERMAN CAMERALISM AS SCIENCE AND PRACTICE

Andre Wakefield

The University of Chicago Press CHICAGO & LONDON

ANDRE WAKEFIELD is associate professor of history at Pitzer College in Claremont, California.

The University of Chicago Press, Chicago 60637
The University of Chicago Press, Ltd., London

Printed in the United States of America

17 16 15 14 13 12 11 10 09   1 2 3 4 5

ISBN-13: 978-0-226-87020-5 (cloth)
ISBN-10: 0-226-87020-0 (cloth)

Library of Congress Cataloging-in-Publication Data

Wakefield, Andre.
The disordered police state : German cameralism as science and practice / Andre Wakefield.
p. cm.
Includes bibliographical references and index.
ISBN-13: 978-0-226-87020-5 (cloth : alk. paper)
ISBN-10: 0-226-87020-0 (cloth: alk. paper)
1. Mercantile system—Germany—History—17th century. 2. Mercantile system—Germany—History—18th century. I. Title.
HB91.W35 2009
330.15'13094309032—dc22
2008036715

∞ The paper used in this publication meets the minimum requirements of the American National Standard for Information Sciences—Permanence of Paper for Printed Library Materials, ANSI Z39.48-1992.

For my parents and Rebecca.

CONTENTS

ACKNOWLEDGMENTS

It has taken too many years to finish this book, and I have incurred too many debts along the way. Let me begin by thanking the teachers, archivists, librarians, friends, and kind strangers who have helped me along the way. You are not forgotten.

I owe special thanks to those who guided me through the early years of this project. Robert Richards was an ideal adviser. He not only writes well, he listens too—a rare combination. Lorraine Daston offered incisive criticism and unwavering support from the very beginning. Michael Geyer provided lots of good advice, most of which I failed to understand until later. Jan Goldstein was always a generous critic. Peter Novick eviscerated my bad ideas in his smoke-filled office. Last not least, as the Germans like to say, I owe much to William Clark and Martin Gierl, who introduced me to the mysteries of Göttingen. Speaking of Göttingen, what would the library cafeteria have been without Gabriel Finkelstein and John Holloran? (Not much.)

Jim Bjork and Tom "Farnsworth" Miles, who were there from the beginning, have seen the good, the bad, and the ugly. Later, at M.I.T.'s Dibner Institute, I was lucky enough to share a building with Len "Boss" Rosenband and Will "Slipper" Ashworth. Both of them have had a profound influence on this book. I am also grateful for the support of colleagues and students in Claremont, California, among them Pamela Smith (now at Columbia University), Judy Grabiner, Richard Olson, George Gorse, Stu McConnell, Dan Segal, Carina Johnson, and Carrie Marsh. Thanks also to Hans Hofmann, Hans-Jörg Ruge, Martin

Rudwick, Ernie Hamm, Alix Cooper, Heikki Lempa, Paul Steege, Jessica "Bobo" Wakefield, and Claudine Cohen. To the anonymous reviewers: thank you for your comments; they were a great help. I have been lucky with my editors: Christie Henry and Karen Merikangas Darling. Nicholas Murray and Adam Rosenzweig helped me see things through to the end.

I am grateful to the Fishbein Center at the University of Chicago, the Max-Planck Institute For the History of Science in Berlin, the (now sadly defunct) Dibner Institute at M.I.T., the German Academic Exchange (DAAD), the Thyssen Foundation and the University of Erfurt, and Pitzer College for institutional support.

Thank you Zachary and Eli, my very own bad cameralists, for keeping dad fresh. Thank you Rebecca for your love and patience. Thank you to my parents for teaching me what I know.

CHAPTER ONE

Bad Cameralists and Disordered Police States

It should be crystal clear, then, that cameralism is nothing more than the science of filling the Kammer with as much money and cash value as the needs of the lord demand. The cameralist, therefore, is a person who procures these proceeds for the Kammer.

—[Maria Machiavel], *Der volkommene Kameraliste* (1764)

They strangled him in a great iron cage—Jew Süss, bloodsucker of the people. Carl Alexander, the Catholic Duke of Württemberg, was dead; now the people were rid of his financier, Joseph Süss Oppenheimer. On his execution day, 4 February 1738, the shops were closed, weddings were delayed, and twelve thousand people lined the streets of Stuttgart, drinking and celebrating. Twelve hundred militia guarded the marketplace, another six hundred surrounded the execution ground, and six hundred more lay in reserve. Dozens of locksmiths and carpenters worked for days to build the killing machine, which featured a scarlet-red cage hanging high above the huge iron gallows. It was an elaborate spectacle. After strangling Süss in the cage, the hangmen hoisted his body up by the neck, so that the people could see it. The corpse, dressed in Süss's signature scarlet red jacket, dangled there, forty feet above the ground—on the tallest gallows in the Holy Roman Empire—as a reminder and a symbol. The body hung there, high above Stuttgart, for six years.[1]

Süss has become a prominent emblem of the anti-Semitism that raged through early modern Germany.[2] But Süss Oppenheimer was not just any Jew. He was privy finance councillor for a Catholic Duke in a Protestant land, where he introduced controversial measures for collecting the state revenue.

Süss helped fix the sovereign finances, and he enraged the estates of Württemberg in the process. After his death, Süss became a symbol of the bad fiscal official—robber of the people, creature of the prince, embezzler of state funds, Court Jew.

I am not interested in the usual questions. Was Süss guilty? Was he innocent? What is certain is that his case resonated through the courts and coffee shops of the empire for decades. People talked about it; people wrote about it. And though his case was remarkable, it was not the only one. Other prominent ministers and officials, like Saxony's Count Heinrich von Brühl or Hessen-Kassel's Wolfgang Günther, were long the stuff of satire and legend, flesh-and-blood symbols of the "faithless servant" or "bad official." There was a whole literature devoted to corrupt officials—bad leaseholders, crooked excise officers, rotten ministers—and though one often spoke about dishonest officials in general, figures like Süss added life and specificity to abstract categories.

Historians, of course, use categories to organize and arrange the motley past. It is what we do. Sometimes we use the language and concepts of the past to arrange our histories, hoping to avoid the myriad traps of anachronism. But it is not that easy, because we inherit words stripped of their symbolic resonances, phrases emptied of specificity and meaning; and yet, we also inherit language that has been reconstructed and remade over time, tailored to fit the needs of changing circumstances. It is a mess.

Twenty-five years after Süss was executed in Stuttgart, an anonymous satirist suggested that his case should serve as a lesson to "cameralists." Writing under the pseudonym Maria Machiavel, the author urged caution. After cheating and bilking the people, cameralists like Süss could not rely on the protection of lords and princes. Careless officials would taste the vengeance of the land. "The Finance Councilor Süs [*sic*], a martyr of the Württemberg *Kammer*," explained Machiavel, "can serve as an example."[3]

Sometimes, things don't make sense. You encounter a trace of the past—a document, a case, a pamphlet—that seems to contradict the terms of the present. This is such a case. Today cameralists are mostly regarded as the propagators of an economic theory called *cameralism*. A quick review of the secondary literature reveals that cameralists were writers and theorists, and that they produced a body of literature that came to be known as the *cameral sciences*. But if cameralists were just early modern economic theorists and writers, what can they possibly have to do with Süss Oppenheimer, martyr of the *Kammer?* That is my question.

Palgrave's *Dictionary of Economics* defines cameralism as "a version of mercantilism, taught and practiced in the German principalities (*Kleinstaaten*) in the 17th and 18th centuries."[4] Cameralists were its propagators. Like many definitions, this one seems relatively straightforward and unobjectionable until you think about it. We have, on the one hand, something that can be taught, a doctrine; we have, on the other hand, something that is done, a practice. It remains unclear how, or whether, the doctrine and the practice are related. We should not blame the messenger. The article merely reflects the latent tensions of a deeply divided historiography.[5] At the heart of the problem lie unresolved issues of science and practice, and the essential question is simply this: What relationship did the cameral sciences bear to administrative practice in the fiscal bureaus and collegia of the Holy Roman Empire? Some have argued that the cameral sciences mirrored administrative practice in the bureaus; others have denied any relationship between the two.[6]

Albion Small's 1909 work, *The Cameralists,* broke with those who had treated the cameral sciences as an early modern economic doctrine.[7] In his opinion historians of economics like Wilhelm Roscher had been guilty of gross anachronism, imposing their nineteenth-century categories and expectations on the early modern map of knowledge. "Cameralism was an administrative technology," he insisted. "It was not an inquiry into the abstract principles of wealth, in the Smithian sense."[8] According to Small, "the cameralists were not primarily economists. They were primarily political scientists."[9] He complained that Roscher had completely ignored the context of early modern administration, thereby neglecting the main import of the cameralist literature. The point, that is, was not economic theory but rather how to ensure a regular supply of money for the prince. The cameralists, therefore, were both writers and administrators, and their works ought to be understood as a kind of how-to literature for aspiring bureaucrats. Small's thesis was endorsed and elaborated by Kurt Zielenziger a few years later, in 1914. Like Small, Zielenziger insisted that the context of fiscal administration was absolutely central to the cameralist literature.[10] The work of Small and Zielenziger would dominate scholarly debate on the subject for decades to come.[11]

In the years before World War I, the fight over cameralism was a fight over disciplinary authority, as sociology sought to limit the jurisdiction of economics over political and social phenomena.[12] At issue was disciplinary identity: to whom did the cameralists belong? For Small, the answer was clear: they were not economists, but men with a normative vision of good government and the good society.

> The cameralists of the books, as distinguished from the cameralists of the bureaus, although the former class was usually recruited from the latter, were the men who worked out for publication, and especially for pedagogical purposes, the system of procedure in accordance with which German governments were supposed to perform their tasks. As a rule these men were employed in administrative positions of some sort, and spoke to a certain extent from experience. They were not mere academic theorists.[13]

Zielenziger too drew a distinction between the "cameralists of the bureaus" and the "cameralists of the books." The former he called "fiscalists," and the latter "cameralists proper."[14] Scientific cameralism, he argued, had arrived only after 1727, with the establishment of academic chairs in the subject at Prussian universities in Halle and Frankfurt an der Oder.[15] The cameralists, in other words, were not misguided economic theorists; they were the writer-administrators and academics who had provided a blueprint for governance in early modern Germany.

In his 1988 book, *Governing Economy,* Keith Tribe fundamentally shifted the terms of this old debate. Drawing on a wealth of published material, he broke decisively with the tradition established by Small and Zielenziger, separating cameralists entirely from the context of state administration. Defining cameralism as "a university science" and placing great weight on the context and practice of university instruction, Tribe divorced cameralism from administrative practice in the Kammer. He claimed, for example, that "after reading several hundred of these texts, I am none the wiser about the organization of an eighteenth-century domain office, the relevant spheres of responsibility, or the proper conduct of account-books." The context of pedagogy, he argued, was much more important. "The two prime influences on these texts were the actual teaching situation and the Wolffian philosophy which informed their style and was itself very largely a product of pedagogic practice."[16] Of paramount importance, then, were the "discursive conditions" under which the texts were produced. In other words, Wolffian philosophy and the seminar rooms of German universities provided the proper contexts for understanding the cameralist literature. Tribe's approach represented a radical departure from existing scholarship on the subject, for he treated cameralism as a self-contained academic discourse and separated cameralists completely from the context of the Kammer.[17]

Tribe's work posed a direct challenge to a century of scholarship. Were the cameral sciences a reflection of administrative practice in the German principalities? Or was there no necessary connection between these sciences and everyday administration in the Kammer? It is another version of the debate that preoccupied proponents of the cameral sciences and their adversaries two hundred and fifty years ago. At that time, academic cameralists were arguing that the business of the Kammer could be reduced to a science, complete with first principles. Their opponents ridiculed the idea, claiming that academic knowledge had nothing to do with practical administration.

It may be time to rethink *cameralism*, a term defined and redefined to fight the disciplinary battles of the nineteenth and twentieth centuries. Economists, sociologists, political scientists, and even public administrators have tried to claim cameralists as their disciplinary ancestors. But the obsession with disciplinary genealogy has obscured the obvious: cameralists, as members of an early modern professional class, are dead. Because cameralists are extinct, they have no modern disciplinary or professional constituency, so we have tried to absorb them into our existing frameworks of knowledge. We have sliced and diced them until they fit our narratives about state building, political economy, and science. But what would it mean to understand cameralists on their own terms? How did contemporaries speak about them? How did they speak about themselves?

Today the cameralist generally appears in our histories as an appendage of something called "cameralism," a disembodied collection of economic principles, academic sciences, and bureaucratic practices. But flesh-and-blood cameralists arrived on the scene well before the more abstract "cameral sciences." Early references to "cameralists," those who managed the prince's finances, had mostly negative connotations.[18] In the late seventeenth century these servants of the Kammer, rapacious bleeders of the people, already had a bad reputation, and cameralists thus became associated with archetypal "bad officials" like Süss.[19] The good cameralist, that utopian servant of the Kammer and protagonist of the cameral sciences, driven only by selfless dedication to the happiness of the people, arrived later. He was a literary and academic invention, designed to improve the poor reputation of the Kammer and its officials. In this sense, the cameral sciences functioned as an academic advertisement for the fiscal policies of the empire's territorial rulers: good publicity.

But there was an alternative to the doctrine of the good cameralist. Satirists, state officials, and academic cameralists wrote constantly about the bad official, that bloodsucker of the commonwealth who, looking only to his own interest, stole money from prince and people. We have long approached the cameral sciences as if they constituted a self-sufficient body of discourse. It was not so. The sciences of the good cameralist were continually in dialogue with narratives about the bad official.

BAD CAMERALISTS

Wilhelm Freiherr von Schröder wrote some unflattering things about "*Kammeralisten*" in his canonical cameralist text *Fürstliche Schatz- und Rentkammer.* Cameralists, he argued, destroyed the foundations of the sovereign income, like those who stripped the forests bare or emptied fish ponds; they were like wild pigs in a garden. Bad cameralists reaped without sowing. There were also economizing cameralists who told the prince how to spend his money. They wanted to limit the money wasted on hunts, performances, mistresses, and other courtly amusements. But princes, explained Schröder, were not like the rest of us; they needed their diversions. Any "prudent and upright cameralist" understood the dangers of excessive economy, which could suffocate a prince's credit and reputation. Even more damaging, though, was the constant fiscal maneuvering and innovation—imposts, sale of offices, monopolies, and the like. These tricks had made *Kammeralisten* universally "hated" and "suspect."[20]

The real problem, as Schröder saw it, was that the Kammer had to oversee "not only expenditure, but also the augmentation and improvement of the revenue." Those cameralists who dealt with expenditures, narrow-minded and penny-pinching, were wholly unsuited to the risky art of creating wealth. Every Kammer needed its bean counters. "But the increase of sovereign income demands completely different people." So Schröder imagined a second body of fiscal councillors devoted to the augmentation of revenue. This branch of the Kammer "would have to be composed of the most subtle geniuses" culled from all provinces of the land, experienced and learned, capable of discerning good projects from bad. These cameralists, salaried projectors of state, would enjoy large incomes and be answerable only to the prince.[21]

Schröder's book was a professional project in its own right, an effort to create a position for himself in the emperor's *Hofkammer.* In fact, many prominent cameralist texts—if not most of them—were strategic and performative; that is, cameralist writings often sought to procure specific advantages, like positions or patents, and they sought to present the author as a good official, the right kind of cameralist. Canonical cameralist authors, from Veit Ludwig von Seckendorff and Johann Joachim Becher to Johann von Justi, secured specific positions through the force of their writings. They argued everywhere and always for employing learned reformers—people like themselves—as state officials.

Becher also wrote about "*Cameralisten.*"[22] "The whole world," he reported in 1668, "complains that cameralists cheat their lords." Moreover, "the old song that cameralists pay nobody, collect a lot, and keep most of it for themselves," made it clear what people thought about cameralists. For Becher the problem was rooted in the four great enemies of the Kammer: "ignorance, indolence, disorder, and dishonesty."[23] One needed knowledgeable and experienced cameralists to combat ignorance. Like every art or science, Becher explained, the fiscal system (*Cameral-Wesen*) had its own "principles, foundations, and axioms," which had to be learned in theory *and* in practice. Skilled cameralists needed knowledge about many things—agriculture, architecture, minting and mining, surveying, water wheels, cattle-breeding, and the law. The good cameralist did not merely procure money for his lord; rather, he managed the finances in a way that served the true interests of the prince. Indolence, the second enemy of the Kammer, could be combated only through constant vigilance. Visitations, audits, and fiscal inquisitions would help to ensure industriousness. Disorder, on the other hand, was a more systematic and insidious evil, revealing itself in everything from cluttered archives to untidy account books. Creeping disorder, the nemesis of accountability, made it impossible to police the Kammer. There could be no order without good manuals, registers, and inventories.

Most dangerous of all, however, were dishonesty and faithlessness. Cameralists, with their proximity to the prince and his treasure, had ample opportunity to enrich themselves at the expense of sovereign and people. Once in positions of power, bad cameralists could act with near impunity, bribing or threatening underlings, making alliances, and hatching schemes. It was important, therefore, to choose one's cameralists carefully. But finding good cameralists was no easy matter. They might be rich without being clever, clever without being honest, or learned without being reasonable.

Becher proposed a solution. By separating the various functions of the Kammer into discrete councils or bureaus, one could create a fiscal machinery largely impervious to the vicissitudes of ignorance, indolence, disorder, and dishonesty. One bureau would manage the prince's capital, another the income, a third the expenditures, and a fourth the examination of claims and demands. Finally, there would be a "*Fiscal-Cammer*" dedicated to the increase of revenues. In short, Becher proposed to banish disorder by restricting communication. If one could make behavior transparent, like a good book of accounts, perhaps the well-ordered Kammer was within reach.[24]

Both Becher and Schröder acknowledged the problem of corruption in the Kammer and suggested ways to improve the situation. It was precisely in this context that they discussed "cameralists." If one could train or recruit better cameralists—less ignorant, more industrious, less haphazard, and more honest officials—or if one could find a way to discipline them, then the sovereign revenue would benefit. It was, Becher argued, *the* most important thing: without ready money, the prince had no authority, and without authority, the people would have neither security nor justice.

The specter of the bad cameralist haunted the cameral sciences throughout the eighteenth century as well, even as the new profession established itself at universities and academies after 1727. Johann Heinrich Gottlob von Justi, the most successful propagator of these new academic sciences, wrote and lectured about cameralist identity in Vienna and Göttingen, and he modeled it too. Justi, who gave himself a noble title, came blowing into Göttingen brandishing a title: Mining Councillor (*Bergrath*) von Justi. Upon arrival, he secured himself another title: chief police commissioner (*Oberpoliceycommissar*). Once in Göttingen, Justi worked tirelessly to protect his honor and his titles in numerous fights with the locals.[25] The chief police commissioner also carried this attitude into Göttingen's lecture rooms. Promising to avoid the dry, pedantic, Wolffian stuff served up at most universities, he proposed something more tasteful and gallant. "I have employed a special teaching style in my courses," he explained. Instead of reading from a textbook for the whole period, he lectured for only half an hour. "Then I got down from my lectern and, standing together with my listeners, I spent the rest of the hour in free and sociable conversation about the lecture material." In this way Justi tried to replicate polite court culture in the classroom. The pedantic professor, so reviled and ridiculed by ministers and state servants, would be replaced by the gallant scholar who could offer something useful and palatable to his students. In short, Justi hoped to infuse the cameral sciences with the grace and style of the princely court.[26]

Justi's classroom lectures and academic writings were at the same time professional projects. Where Schröder had hoped to create a special council of fiscal geniuses in the Kammer and to secure himself a position in it, Justi proposed an entirely new "faculty of œconomic and cameral sciences." Cameralists would join theologians, physicians, and jurists as a new breed of university-trained professional. In addition to lecturing, they would offer advice about state projects and initiatives, much like Schröder's geniuses of the Kammer. They would train good cameralists for the forests, mines, domain offices, and treasuries of the empire.[27]

But academic cameralists like Justi, purporting to offer public lectures about the prince's most intimate affairs, were in a precarious position. They were the public face of secret things, presenting the fiscal-police state as it ought to be. Good cameralists, the mythical officials of this well-ordered police state, collected revenue only from legitimate sources, without infringing on the rights and privileges of the estates or the people. Academic cameralists like Justi would teach "future cameralists" how to be good cameralists. Because good cameralists collected revenue only from the prince's own possessions and ordinary revenues, lectures focused on the regalian rights and sovereign possessions—mines, forests, and domain lands; because good cameralists collected taxes willingly contributed by the estates and the people, lectures would focus on how to collect revenue without overburdening the populace; and because the prince could only increase his revenue by increasing the wealth of the land, lectures would demonstrate how to improve existing sources of revenue and find new ones, according to the dictates of "good police."[28] To serve the good cameralists of the future, then, academic cameralists promised to reduce the seemingly haphazard practices of the Kammer to a science. They would enlist natural philosophy and natural history in the service of "good œconomy" and secure the foundations of the police state with the latest philosophical principles.

But good cameralists had a bad reputation. Skeptics and satirists suggested that academic cameralists like Justi—the writers, professors, and theorists of cameral sciences—were either incompetent or dishonest. Some even suggested that a science of the Kammer was impossible. Cameralists were simply artisans who learned their jobs by routine and memorization, like shoemakers or carpenters. There were also those who saw something more sinister in the cameral sciences, which they regarded as propaganda for finance ministers determined to fleece the people. These skeptical and critical voices have been silenced in the secondary literature on cameralism, which mostly reflects and analyzes the well-policed reveries of academic cameralists. This may explain why it often feels like one side of a conversation. We have inherited the utopia without the dystopia, the ideology without the criticism, the good cameralist without the bad cameralist.

Broadly speaking there were two sorts of attacks on the cameral sciences. Some critics ridiculed academic cameralism as "impractical," arguing that it was of no use to anybody. Others saw something more cynical in all the talk about general welfare, paternal princes, and good police. For them, the cameral sciences served the treasury and *not* the interests of the people; this implied, in turn, that the interests of the prince were different from those of his subjects. It was explosive stuff, and those who dared to write it had to hide behind satire and pseudonyms.

"Maria Machiavel," whose 1764 work *Der volkommene Kameraliste* attacked cameralists and their sciences, was one of these pseudonymous satirists.[29] "My ancestor wrote practically," explained the author; "his great opponent refutes him theoretically." This "great opponent" was Frederick the Great, whose *Anti-Machiavel* (1740) had directly attacked Machiavelli's *The Prince,* arguing that a king's true interest was the welfare of his people.[30] "Our cameralist writings are of the same kind," he continued. "They teach theory; I show practice."[31] The primary target of Machiavel's satire, however, was not Frederick II, but Justi. "What is cameralism [*das Kamerale*]?" asked the author. "Is it a science? Is it an art? Wherein consists the art or science of the same?" he asked, mimicking the preface to Justi's *Staatswirthschaft.* "Should I give the explanation that one finds in the cameralist writings?" Machiavel continued. "Everyone knows it, and everyone knows that it's not right."[32]

The science of the Kammer was a chimera, Machiavel went on to explain. Certain speculative types, mostly professors, claimed that cameralism was a true science with its own unshakeable foundations and principles; they hoped to embarrass practicing cameralists, who conducted the business of the Kammer mechanically. But did anyone really believe that cameralism was like theology or philosophy? "Who learns its first principles? Who teaches them? And how necessary is it in order to become a good cameralist? Not any more necessary than it is for learning a trade. One only has to know the sources from which to draw money; one only has to know a few formulas." Cameralism, the business of the Kammer, was a mechanical art, demanding only memory and repetition. The cameral sciences were a sham.[33]

Experience, continued Machiavel, clearly demonstrated that skilled cameralists were more like shoemakers than philosophers. "One can see that pure coincidence is enough to make the most perfect cameralist out of a scribe, cook, hunter, etc. without any prior study." There were plenty of examples. "A. was the court fisherman. His skill at fishing gave him access to the forest office; it wasn't long before he was in the *Kammer* fishing for money, instead of in the streams catching fish." In this way, cooks, pages, messengers, and other subalterns had insinuated themselves into the Kammer.

Every prince, he explained, preferred clever schemers to scholars of economy and good police. Princes needed their cameralists to find money. The job description was simple: meet the financial needs of the ruler. Nothing more. All the talk about good budgeting and economy was so much bunk. Did anyone really believe that the servant should scold the master? Could anyone be stupid enough to tell a prince how to spend his money, or how to economize? Questions like these were not for cameralists, but for moral philosophers, and "we do not want to dwell on this gloomy and sad doctrine, because it is so hated at most courts and *Kammern,* and it cannot be applied."

According to Machiavel, the prince's Kammer had three main potential sources of revenue: (1) the people; (2) the regalia of the sovereign (like special mineral and forest rights); and (3) the domain lands. Though textbook cameralists liked to focus on the regalia and domains, "these are usually very dry." Instead, every skilled cameralist knew that one had to invent extraordinary devices for collecting money, because the ordinary revenues were always insufficient. "One must treat the subject like a sponge."[34]

For Machiavel, the well-ordered and beneficent police state was a fantasy. "No one really believes," he explained, "that the sovereign's main concern is the welfare of his subjects."[35] The territorial princes of the empire had been trampling on the traditional rights of their subjects for decades, a trend that had climaxed during the recently concluded Seven Years' War (1756–1763). Cameralists were the experts at deception, constantly developing new techniques to exploit the people. Though some stubborn souls in Germany and Italy insisted on their ancient rights and freedoms, the wise cameralist could always invent clever techniques to get around them.

> He must work very quietly in these matters. He must be like a miner who works so secretly that one is unaware of the subterranean fortifications until it is too late, so that the people cannot hinder and destroy them. In short, he has to extract a little here and a little there under all kinds of artful and sweet-sounding pretenses. The excise, monopolies, fines, stamped papers, surveying, and a hundred more things like this offer him objects upon which he can exercise and sharpen his wit and inventiveness.[36]

Within this secret world of fiscal inventiveness and pretense, academics had their own special role as propagandists for the fiscal-police state. Masquerading as friends of the people, they justified the rapaciousness of princes. There were many ways for a cameralist to "enrich the *Kammer* under cover of the general good."[37]

Machiavel's satire may have been funny, but it was deadly serious. Cameral science was a dishonest enterprise. For all their blustering about the common good, cameralists were creatures of the prince; for all their treatises about sustainable forests and happy peasants, cameralists earned their pay as public relations men for the sovereign. How, after all, could one have an honest public literature about the most secret affairs of the king? Impossible. As the public voice of the fiscal-police state, the academic cameralist had to sell the policies of his sovereign. That was his real job.

Anonymous satirists, however, were not the only ones to obsess over bad fiscal officials. Johann von Justi's books are filled with admonitions about the importance of integrity and the dangers posed by dishonest officials.[38] He also published a stinging criticism of Count Heinrich von Brühl, prime minister to the elector of Saxony, Frederick August I, sometime king of Poland.[39] The book was very successful, appearing in English translation during his lifetime. Justi's treatment, which immediately became the standard, has defined Brühl's administration ever since.[40] The count, it argued, had single-handedly bankrupted Saxony, the empire's most industrious and bountiful territory. Starting as a lowly page, Brühl had managed to secure the favor of August the Strong. Soon he showed a gift for raising money. He was, argued Justi, a ruthless "bloodsucker," a creature of the elector, and a self-interested embezzler of public moneys.[41] He was, in other words, the archetypal bad cameralist.

Among Brühl's many crimes, argued Justi, were capital offenses. "It is a standing maxim, allowed by all civilians," he explained, "that the domains of electorates and principalities cannot be legally alienated." And yet this was precisely what Brühl had done, taking some of the elector's domains into his possession, thereby breaking his oath as president of the treasury. Moreover, Justi made it clear that Brühl was a *type:* the "regent-minister." He was the "grand-vizier" of Saxony, oppressing the land like an Ottoman despot and trampling on the rights and liberties of the people. Perhaps most important, Brühl was uneducated. He had risen to his position through intrigue, luck and flattery. Brühl's case, though dramatic, was by no means unique. "Believe me," continued Justi, "I have seen ministers of commerce unacquainted with the very first rudiments of trade; I have known a president of mines who did not know common lead ore; ministers of finances without any notion about the purpose of the financial system."[42]

Justi's attack on Brühl can be read as a companion piece to his earlier *Staatswirthschaft,* whose preface included an extended pitch for academically trained cameralists. Justi might well have been referring to Brühl, for example, when he wrote that "in various lands those who now hold the most prominent state offices were once lackeys, runners, scribes, common hunters, small-time collectors, and the like." This in turn had led to disorder and corruption, because the practical cameralist lacked "a philosophical head, the accompanying insight into the whole, and the gift of coherence and good categories."[43] In fact, Justi suggested that science (*Wissenschaft*) itself might protect against the favoritism and corruption of the courts.[44] "In general," he explained, "those cameralists who have merely been reared in the affairs of state almost all share the failing that they are too much beholden to the interest of their lord."[45] Seeking only to please the prince, practical cameralists neglected his real interest, thereby sacrificing the welfare of the land and the happiness of the people. Universal cameralists, armed with coherent axioms and well-ordered principles, would make no such mistakes. The cameral sciences, and the universal cameralists who had mastered them, would serve the true interests of people and prince. That was the cameralist dream, a dream that derived its power from the cameralist nightmare of fraud, corruption, and disorder.

THE DISORDERED POLICE STATE

Historians like to have their epiphanies in archives and old libraries. Some people make fun of this. How naïve, they say, to imagine that truth waits in some fascicle or book, waiting to be discovered; how silly to fetishize the archive. They are not altogether wrong. Even those of us who do not believe that truth sits on the shelf, waiting to be discovered, sometimes act as if it does. More often than not, though, archival finds confirm what we already know. That does not make them any less valuable.

I had one such revelation in a seventeenth-century palace, a massive baroque building erected by Ernst the Pious after the Thirty Years' War. There, in the old library, I opened an obscure, anonymously authored, and seemingly innocuous book about excise taxes, and out fluttered a letter. Addressed to Prince Friedrich II of Sachsen-Gotha, and dated 7 April 1717, the letter had been sent from Halle by someone named Happe.[46] "Serene Duke, graciously reigning Prince and Lord," it began, "if one wants to see the model of a God-fearing, upright, and perfect prince, who loves all his subjects like his own children, and who works gladly for their perpetual welfare and increase, then it is surely your Serene Highness." Happe, who was sending his book along to Prince Friedrich, marked "certain important passages, where they begin and end, with horizontal lines," so that the prince could go through them in "half an hour." Flipping through the book that day, I indeed found little dashes penciled in the margins. Happe, it appears, wanted to highlight the interconnected problems of fraud, useless ordinances, and faithless officials. He also wanted a job.[47]

The plague, Happe claimed, had been mere child's play compared to the onslaught of police ordinances, edicts, statutes, and orders issuing from the many chanceries, chambers and councils of the Holy Roman Empire. All remedies seemed powerless to stop the madness, and no "state physician" had yet found a cure. The ordinances published during the last generation alone, Happe continued, could fill St. Peter's Basilica in Rome, and things seemed to be deteriorating by the day. "Hasn't it constantly gotten worse and more vexing from day to day? Haven't all these measures enjoyed as little respect and produced as little as empty paper, with nothing written upon it? One has treated them like bits of chaff or worthless slags." Happe considered the situation serious enough to merit radical therapy. Like a poisonous weed, the existing system, a system based on lies, had to be extirpated. Instead of publishing more edicts and ordinances, which nobody observed anyway, it was time to think about enforcement. Happe imagined himself an inventor-entrepreneur of state finance, a Leibniz of the treasury, who would engineer a better system. And he had the "machine" to do it.

Most palaces, he explained, had empty rooms and apartments that could be used to house all the meetings of the prince's various councils and colleges—the chancery, the fiscal chamber (*Kammer*), and the secret council (*Geheimer Rat*), among others. More important, the building could be configured for surveillance. A network of tunnels leading to a series of cabinets, or "loges," would allow the prince furtively to watch his officials. In each cabinet, the windows would be covered by "very fine lattices, screens and the like, so that the person sitting in the loge, while seeing and hearing what happened and what was said in the chamber, could not himself be seen." Happe imagined cabinets like these in every room, so that the prince could observe the business of state whenever and wherever he pleased. He might even "write, read, eat and drink" in his box, where he would "hear and see everything that was happening there, and no person would know whether or not he was in the loge."

Happe claimed that his machine would "instill great fear in the prince's servants, since they would at all times have to worry that he was sitting in his cabinet." It would benefit the whole land. "There would be such industriousness, order and loyalty!" The invention would work whether or not the prince was observing, since his officials and subjects "would have more fear of the empty cabinet, or loge, than they once had of the prince himself." Moreover, Happe thought that his invention would spill out beyond the palace, into the streets and structures of territorial towns and cities. Every minister, director, nobleman, state official, collector, merchant, banker, artisan, farmer, or burgher could make use of the same principle. Everyone would have the sense of being watched. "All work will be done more honestly and industriously. *Many millions less will be stolen, or can be stolen, than up until now.*"[48] As if to drive home the threat of constant surveillance, Happe suggested that the doors of the palace be emblazoned with a motto: "God sees and hears everything; the emperor, king or prince [sees and hears] as often as he wants."

There are of course echoes—strong echoes—of Bentham and Foucault in all of this, but they are the echoes of the present, not the past. Happe preceded Bentham by almost a century, and though the similarities between them are undeniable, the differences may be more instructive. I have no evidence, for example, that Happe's loge was ever built or that any prince even bothered to read his book. Nor did his panoptical dream aim to discipline prisoners; it was a vision for monitoring fiscal officials. But like many other contemporary projects and plans, this one has a whiff of desperation about it. Impoverished scholars and academics, in search of patronage and preferment, were especially prone to construct elaborate visions of control and order. It would be a mistake to take them too seriously.

Happe's panoptical vision may be less significant than his claims about the police ordinances, which undergird the historiographical foundations of the "well-ordered police state." Ever since Marc Raeff used the phrase in 1975, it has enjoyed a robust afterlife in English-language books and journals.[49] Raeff argued that "within the constricted framework of the middling and petty states of Germany" the mania for regulation "easily led to the tyrannical control and supervision of every facet of public and economic life." Relying on the *Landes- und Polizeiordnungen*—statutes, orders, edicts, police ordinances, and the like—as evidence, Raeff argued that "cameralist police and mercantilist economic policies" had successfully created modern, resource-maximizing societies in the principalities of the Holy Roman Empire.[50] It was a powerfully persuasive and original argument, one that shifted the locus of modernity from eighteenth-century Paris to seventeenth-century Germany. In making it, Raeff assumed that these ordinances (in contrast to "theoretical writings, plans and projects") reflected administrative practice on the ground.[51] In other words, he assumed that they were actually enforced, and that they *shaped* the everyday existence of target populations. Some commentators recognized that this assumption was open to debate, as did Raeff himself. Writing in 1978, for example, Mack Walker cautioned that Raeff "should be careful not to assume too close a relation between doctrine or legislation and material effect: but there is vast uncertain room for debate about this."[52]

The tremendous success of Raeff's essay has served to obscure the evidential basis upon which it rests. Over time the "well-ordered police state" has assumed the patina of established fact, so that it now sometimes appears as a disembodied, stand-alone, self-evident phrase.[53] It is also one of those historical constructs that has "jumped species," finding a welcome in other disciplines like political science and anthropology.[54] Moreover, the concept has been folded into a growing body of literature that stresses the progressive disciplining of society.[55]

If we are to believe Happe, however, the "well-ordered police state" was nothing more than a paper tiger, existing only in the dreams and delusions of the statute writers. In theory it was a beautiful system; in practice it was an empty promise. Looking back from the present, historians have discerned a rising tide of discipline during the seventeenth and eighteenth centuries; looking forward from the past, Happe and his contemporaries were more likely to sing the anxieties of a recalcitrant and disordered world.

CAMERALISTS AND THE KAMMER

The Kammer, as Happe's dream of surveillance suggests, was more than some abstract fiscal-juridical concept. It was a physical space, a chamber where fiscal officials met to discuss the most secret affairs of the prince. The character of the Kammer—its size, its business, its members—varied dramatically from territory to territory. Smaller principalities, like Sachsen-Gotha, might have a small Kammer, staffed by a few officials. Others, like Prussia, had elaborate systems of fiscal administration, with a central organ, like the General Directory, and many subsidiary provincial Kammern. By the early eighteenth century most territories in the empire had a Kammer. It might be big or small. It might generate most of its revenue from silver mines, salt, forests, beer, crown lands, or tolls. But in almost every case it was called the Kammer, a chamber where fiscal officials debated important issues and voted on them.[56] From this central chamber, the Kammer extended its reach outward to the farms, forests, and mines of the periphery. These "ordinary" sources of sovereign revenue constituted the symbolic heart of the Kammer. The good prince of the cameral sciences, unwilling to burden his subjects with oppressive taxes and contributions, knew how to raise enough revenue from his own lands and regalian rights.

The origin of the Kammer as a specialized, collegial body dedicated to administering the sovereign finances has been the subject of considerable debate.[57] Those towering administrative historians of the late nineteenth and early twentieth centuries, scholars like Eduard Rosenthal and Otto Hintze, largely disregarded the Kammer as they traced the development of the secret council (*Geheimer Rat*).[58] Their followers, however, discovered its significance. Melle Klinkenborg identified the Kammer as a key site of fiscal-political direction in Brandenburg.[59] Gerhard Oestreich and Werner Ohnsorge then argued that Klinkenborg's thesis held true for other German territories as well.[60] The Kammer, they claimed, had provided a basis for many of the secret councils that arose during the seventeenth and eighteenth centuries. Surveying the literature in 1962, Ulrich Heß argued that a new "secret sphere" of sovereign governance, centering around the princely finances, had evolved in the Kammer during the second half of the sixteenth century.[61] By the seventeenth century most German territories, large and small, had developed Kammern to manage the intimate affairs of princes, dukes, kings, and emperors. By the second half of the seventeenth century, members of the Kammer began to be recognized as a distinct group. People started calling them cameralists.

Every cameralist text looked to the Kammer. All that talk about mines and forests, minerals and manure—usually ignored or passed over in the secondary literature—represented more than some technical or scientific curiosity; it was also a moral gesture designed to demonstrate the attributes of the good cameralist. Every responsible fiscal official, that is, had to know his way around a mine or a barley field, because those were the appropriate "ordinary" sources of revenue for his prince. Later, when "aspiring cameralists" flocked to places like Göttingen and Lautern to hear lectures in the cameral sciences, they were not just studying mercantilist policies and the principles of good police. They were learning how to behave as members of the Kammer. It was not enough to know about budgeting and accounting; one had to be fashionable too.

Insofar as cameralists sought to systematize the daily work of fiscal administration, they faced great obstacles, because the logic of every Kammer was distinct, attuned to the local resources of a particular territory or region. The Holy Roman Empire, with its hundreds of kingdoms, duchies, principalities, and bishoprics, presented a staggering diversity of administrative structures, geography, and economic activities. Accordingly, cameralists filled their books with endless detail about everything from pigs and iron mines to forests and barley fields. This has led commentators to suggest that the cameral sciences were descriptive sciences, models of "practical reasoning" that avoided the utopian thinking of nineteenth-century economics.[62] There is only one problem: the forests were idealized, and the pigs were utopian—all fat and stall-fed, with good manure.

Cameralists did not just dream about perfect pigs; they also wrote about utopian principalities. Veit Ludwig von Seckendorff's *Teutscher Fürsten Stat* (1656) portrayed the little principality of Sachsen-Gotha as a model state, and it cast Duke Ernst the Pious in the role of ideal Lutheran prince, a ruler who cared for his subjects as a father cared for his children. Historians have long relied on Seckendorff's *Fürsten Stat* as a kind of ideal type, using it to elucidate the structures of everyday life and social discipline in petty Lutheran principalities.[63] The success of Seckendorff's book transformed Sachsen-Gotha into *the* model principality, complete with compulsory schooling, public health, an orphanage, and lots of moral regulations. Seckendorff had a hand in drafting the police ordinances for Sachsen-Gotha too, and his *Fürsten Stat* looks very much like the regulatory apparatus of well-ordered Gotha.[64] As if to reinforce the point, a statue of pious Ernst, holding a bible, still greets tourists in front of Schloss Friedenstein, the massive palace he built in Gotha after coming to power in 1640. When I visited Friedenstein, I wondered about something else: how did he pay for it all?

The short answer is wood and taxes. Soon after taking over Sachsen-Gotha, the duke established a Kammer to oversee all income and expenses. During the first decade of his rule, Ernst's new Kammer quadrupled the revenue.[65] Despite huge debts from the Thirty Years' War, Ernst moved ahead aggressively with construction of his new palace. By 1654 it was nearly completed.[66] During the early years of his rule, Duke Ernst and his officials in the Kammer behaved just like bad cameralists, looking everywhere for new sources of income. The Kammer imposed an excise tax on grain in Gotha during 1644. Not only were excise taxes highly unorthodox in Ernestine Saxony, but it was especially ruthless to impose a grain tax on hungry people during wartime. And it did not work. Farmers took their grain to "foreign" mills (i.e., those in nearby principalities) to avoid the tax. The Kammer responded by imposing a tax on all grain and flour brought into the city, which made things even worse for its residents. When members of the town council complained that the grain and flour excise was ruining Gotha, the Kammer ignored them. Ironically, Ernst's Kammer proved incapable of enforcing its nasty excise. The duchy was just too small.[67]

Though Gotha's Kammer failed to fleece the people with new taxes and tolls, it had more luck in the duke's forests. Ernst had inherited the most valuable parts of the *Thüringerwald,* large tracts of thickly forested hillsides and valleys that stretched through the storied districts of Tenneberg, Reinhardsbrunn, and Georgenthal. Pious Ernst liked to climb through his forests to the Inselsberg, the highest mountain in Thüringen, and lounge around in his garden house at the top. Young Seckendorff—he was only twenty-three years old—was good at composing cheesy poetry to flatter the duke, who was just like the mountain, mild and omniscient.[68] Braunschweig-Lüneburg's Brocken was higher, and Brandenburg Prussia was much bigger, but on his little mountain Ernst could imagine himself king of the world. After hearing Seckendorff's poetry about him, the duke must have realized that the young man could make anything sound good, even a sad, small, poor, disordered territory like his.

The Kammer, which oversaw management of the duchy's forests, did not have the same luxury. It was constantly on the prowl for new ways to make money. With the rising demand for wood after 1650, and with closer oversight, Ernst's forest officials were able to squeeze more revenue out of his "children." Other projects, however, went very wrong. When the Kammer tried to develop a market in Bremen and Amsterdam for its "mast trees," it badly miscalculated what they were worth.[69] The Kammer also issued an elaborate "forest ordinance" in 1644 that aimed to protect "the land's treasure" for future use. The real purpose of the thing, of course, was to secure more revenue for the duke's treasury by prohibiting various kinds of customary use. This failed too. Despite rising revenues, the duke's forest officials were never able to transform the rollicking *Thüringerwald* into the well-policed forest of cameralist dreams.[70] Seckendorff knew about all of this, because he was in charge of Gotha's Kammer.[71] His *Teutscher Fürsten Stat,* the seminal cameralist text, was also a staggeringly successful piece of propaganda.[72]

When Seckendorff resigned his post eight years later, in 1664, he penned an extraordinary indictment of Duke Ernst and his government: "Causes which move me to leave the court, even though it has brought me temporary honor and pleasure, and even though the change will do not a little harm and inconvenience to my finances." The document was never published, and it sat there, in a family archive, for almost three hundred years before anybody noticed it. Seckendorff hated his responsibilities in the Kammer and the secret council, which "always give one occasion for anger, dissimulation and other troubles, which very much hamper Christian love and contemplation."[73] Some of it was the workload, which he found overwhelming. The real problem, however, was the duke and the lying. Pious Ernst was a screamer who regularly abused his most important officials and ignored their advice. He mistrusted his closest advisors, and he underpaid them. The constant suspicion and complete disorder that plagued daily business at Ernst's court had quite literally made Seckendorff sick. It had gotten so bad that he could no longer help "his honest servants and friends." Adding insult to injury, Seckendorff even disapproved of how Ernst was raising his *own* children, the future princes of

Sachsen-Gotha-Altenburg.[74] Ernst's Gotha, the mythical model police state, was a bad place for good cameralists.[75]

THINGS TO COME

In 1909 Albion Small argued that Seckendorff, who had "systematized Duke Ernst's scheme of life" and "composed Ernst's practices as a manager into a didactic treatise," was the "Adam Smith of cameralism."[76] Small's thesis has lived a healthy afterlife in the secondary literature. Despite Keith Tribe's lonely dissenting voice, cameralism survives in our historiography as a *practical* science that mirrored fiscal practice. Seckendorff's *Fürsten Stat,* built on the well-policed rock of Pious Ernst's mythical Gotha, serves as ground zero for this line of argument; it is, as I have tried to show, a very unstable rock.

If Seckendorff has functioned as the bedrock of our cameralist narratives, then Johann von Justi has been their keystone. It was Justi, argued Small, who shaped "cameralistic technology" into a system of theory. Given that Seckendorff was the Adam Smith of cameralism, he continued, "we may carry out the conceit by calling Justi the John Stuart Mill of the movement."[77] Small, who considered Justi the quintessential cameralist, originally planned to present "Justi alone as the type of cameralism in general."[78] The historiography of cameralism has developed considerably during the last century, but Small's claim still holds today: Johann von Justi appears almost everywhere as the essential and typical representative of German cameralism.[79]

This book takes aim at Seckendorff and Justi, bedrock and keystone of cameralist narratives. By rethinking their published works in the context of their professional ambitions, I want to suggest a different narrative about the cameral sciences. I hope to show that we cannot read these texts in isolation, as a self-contained body of literature that makes sense on its own. I hope to demonstrate that the cameral sciences were intimately and ineluctably tied to the sciences of nature, especially chemistry and mineralogy. I hope to convince you that these "practical sciences" were, in their way, just as idealistic and romantic as Novalis's blue flower or Young Werther's yellow vest. Like the romantic heroes of a later age, Justi's good cameralists longed for better worlds—clean streets, healthy cattle, happy people, beneficent nature—in the face of disorder and poverty. These cameralist dreams often arose out of everyday frustration with silver mines and fulling mills, model farms and œconomic gardens, useless professors and unruly students.

But this book is not just about cameralists and the Kammer. It is also about the fiscal landscapes of the German-speaking lands of the post-Westphalian

Holy Roman Empire. That is a nasty mouthful, and I do not want to repeat it, so please forgive me if I call the thing "Germany." Contemporaries, including many of those who appear in these pages, referred frequently and unproblematically to *Deutschland.* That does not mean they knew what they were talking about. As Rudolf Vierhaus once pointed out, it can be "difficult to ascertain what constituted "Germany" in the seventeenth and eighteenth centuries."[80] And the more you think about it, the harder it gets. The empire was a thing without boundaries—not really a state, not entirely "German," and without clear lines of authority. For nineteenth-century historians, especially Prussian historians, this made it weak, suspect, and undesirable, a thing to be overcome. That may have been true enough for larger states like Prussia and Austria after 1648. But for smaller entities the empire served as a kind of "incubator," protecting delicate political organisms—ecclesiastical principalities, imperial cities, tiny *Reichsgrafschaften*—that otherwise might have been gobbled up by predatory neighbors.[81] The staggering political and fiscal diversity of the Holy Roman Empire included European powers and tiny principalities, silver states and agricultural territories, large administrative departments and one-man bureaus. "Germany" was a mess.

Enter the cameralists. Following the Peace of Westphalia (1648) many of the larger and middling states of the empire became increasingly independent, developing diplomatic and fiscal policies that ran counter to the interests of the *Reich.* Cameralists usually appear in our histories as the shock troops of such territorial state building; they were supposedly the mouthpieces of state formation, giving voice to the practical problems that faced central European bureaucracies after the Thirty Years' War.[82] Cameralists thus assume their role as mirrors of the world, authors whose books reflected the practical experiences of state servants. An unarticulated corollary of this claim suggests that we may use the cameral sciences as evidence of past bureaucratic practice, because they were *descriptive* sciences.[83] That is how idealized texts, originally crafted to please powerful people by sketching well-ordered possible worlds, have come down to us as reflections of administrative practice. If the book does anything, I hope it forces a reconsideration of these sources and how we use them.

This is a book about cameralists and the Kammer, about the public discourse of the cameral sciences and the secret discourse of the fiscal chamber. I explore the relationship between those two worlds in the following chapters by pursuing Johann von Justi from the mines near Vienna, through the university town of Göttingen, and into his Prussian prison cell along the Oder River; by tracking Johann Heinrich Jung-Stilling from his model farm in Siegelbach to the lecture halls in Heidelberg; by following Johann Beckmann from Linnaeus's Uppsala to the œconomic garden in Göttingen; and by shadowing Veit Ludwig von Seckendorff from Pious Ernst's Kammer to the family estate in Meuselwitz. Our cameralist tour will pass through large kingdoms and small duchies, silver mines and forests, farms and ironworks, lecture halls and woolen manufactories. But though the scenery changes, my purpose remains the same: to interrogate the functions of knowledge in the mines, manufactories, forests, farms, ironworks, academies, and universities of the empire's fiscal-police states.

SCIENCE AND STATE BUILDING

In his seminal 1918 article, "The Crisis of the Tax State," Joseph Schumpeter located the origins of fiscal modernity in the transition from "domain state" to "tax state." By harnessing modernity to taxation, he implied that domain states like Sachsen-Gotha, whose finances relied on crown lands and regalian rights, embodied fiscal backwardness.[84] The argument was, as Schumpeter admitted, based primarily on Austria and Germany. Some eight decades later it is just those territories—the German and Austrian lands of the seventeenth and eighteenth centuries—that have caused most trouble for the disciples of Schumpeter's model. Prussia, especially, sustained its substantial army and large administration with a fiscal system that looks suspiciously like a streamlined—dare I say modernized?—"domain state."[85] The rest of the early modern German and Scandinavian lands, usually disregarded in our histories of European state building, also behaved much like modernizing "domain states."[86] Here too, however, presumed fiscal peculiarity is generally explained away as some kind of retrograde aberration.

But there is another kind of narrative about the rise of the modern fiscal state, one epitomized by James C. Scott's *Seeing Like a State.* For Scott, the "fiscal forests" of eighteenth-century Prussia and Saxony provide *the* model of modernity, and Prussia's eighteenth-century administrators, who viewed forests "primarily through the lens of revenue needs," begin to seem more like the proto-managers of a multinational timber corporation than the members of a backward fiscal apparatus. Instead of a domain-state backwater, Prussian forests become an archetype for the "simplification, legibility, and manipulation" so characteristic of the modern state.[87]

My intention here is not to challenge the validity of Schumpeter's model.[88] Nor do I wish to interrogate Scott's picture of forest management. I just want to point out that we have a problem. Maybe the modern state does exhibit a progressive evolution from domain state to tax state, so that the German and Scandinavian lands did lag behind France and Britain. Or maybe the domain lands of eighteenth-century Saxony and Prussia were at the cutting edge of fiscal modernity. But we cannot have it both ways.

Each approach has its merits. Schumpeter's followers, led by Richard Bonney, have given us a much better sense of the revenue sources of early modern states, demonstrating that, despite certain common features of European state building, the *revenue mix* could vary dramatically from state to state. Some, like England/Britain, relied almost completely on customs duties and excise taxes.[89] Others, like Prussia and Saxony, got much more of their funding from domain lands, state-run enterprises, and sovereign privileges (e.g., mineral and forest rights). Developmental models of the fiscal state, regardless of their flaws, have thus performed a great service by encouraging the collection, synthesis, and analysis of much data. Other more qualitative accounts of the modern state, like Scott's, have their advantages too. By focusing on certain features of state activity, like "legibility" or standardization, they sometimes unearth moments of modernity in unexpected places, thereby providing a corrective to sweeping developmental narratives.

But different as they are, these conflicting narratives of the modern state all share one essential attribute: they seek the origins of the present in the past. It is possible to imagine a different kind of fiscal-administrative history, one that looks forward instead of back, taking seriously the dreams and ambitions of the state makers and seeking possibilities rather than origins. Such a history might make the "rise of the modern state" seem a little less inevitable and a little less monolithic; it might conjure a time when the models of state building were many and diverse, and when the road to modernity could have passed through any one of them.

If nothing else, such an approach would allow us, we historians of "backward" states, to focus more on what actually happened and less on what failed to happen.[90] Most existing models of state formation are normative and, despite frequent claims to the contrary, teleological and anachronistic. You know the drill. Start at a given place and time, be it seventeenth-century Holland or twentieth-century America, and ask, how did we get here? These accounts of progressive development imply a correspondingly stagnant world of less successful and less modern alternatives. But backwardness is in the eye of the beholder. Sitting where we do now, able to survey great changes from a distance—the explosive industrialization of the nineteenth century, the transformation of the globe in the twentieth—one is tempted to look back on the relatively quiet lands of the eighteenth century with a kind of patronizing nostalgia. In a world of integrated financial markets, nuclear weapons, and multinational corporations, it seems clear where things were headed. The narcissistic present forever seeks itself in the past.

This has been especially true of administrative history. It is by now a truism, for example, that the *Acta Borussica,* that remarkable monument to the older school of Prussian historiography, was written as the success story of a self-confident, young German nation.[91] Today's *Acta Borussica*s, the many memoirs of Anglo-American power, are similarly the retrospective accounts of our own "rise."[92] What gets neglected, of course, is the "losers' history" of state formation. What about those middling and smaller states of the Holy Roman Empire, the Galapagos Islands of state building? If the history of state finance is about progressive models of development, about the relentless march to *us,* perhaps those territories have nothing to offer. But if we approach that history as a possible source of insight about other things, like culture and knowledge, then the neglect seems unwarranted. If, that is, we suspend judgment about the inevitable destiny of Europe's fiscal dodos, replacing our categories of hierarchy with difference, perhaps we can begin to understand what these revenue systems actually produced instead of how they failed.

Among the things that central Europe's fiscal-police states *did* produce was science. Schumpeter himself, in an uncanny premonition of much recent research, explained that "tax brings money and calculating spirit into corners in which they do not dwell as yet, and thus becomes a formative factor in the very organism which has developed it."[93] As it happens, historians of science have shown great interest in practices of standardization, quantification, and objectivity, and they have linked those practices to *general* patterns of state building and administration.[94] There has, however, been much less research into how *specific* fiscal systems produced certain kinds of knowledge. It may be time to ask, for example, why so many of the most prominent figures in German science and literature—Leibniz, Goethe, Novalis, Abraham Gottlob Werner, and Alexander von Humboldt—were seriously involved with mines and mining; or why "scientific forestry" took root in these same places (Saxony, Hannover, and Prussia); or why alchemy and metallurgical chemistry thrived there too; or why "technology" as an academic subject originated in Hannover.[95] And so on. My point is simply this: the fiscal structure of the German territories, whether backward or not, was a hothouse for certain kinds of knowledge. Silver mining, long a backbone of the state finances in Saxony and Hannover, provided support for a whole array of chemical and earth sciences. State forests and farms provided laboratories for certain kinds of botany and agriculture. German universities and academies, which were also expected to generate revenue for the state, engaged in a grand eighteenth-century scheme to sell the sciences. In a word, the fiscal logic of the Holy Roman Empire, whether backward or not, was certainly not neutral or universal. It was an extractive logic, attuned to the particularity of local places and populations. It was also a creative logic, producing knowledge even as it yielded revenue.

German cameralists existed at the nexus between science and economic development. They were the preeminent German proponents of the notion that one could promote development through systematic application of the natural and human sciences. In this sense their efforts constitute the perfect laboratory to test claims that "scientific culture" paved the way for early industrialization.[96] In Joel Mokyr's formulation, "The Enlightenment in the West is the only intellectual movement in human history that owed its irreversibility to the ability to transform itself into economic growth."[97] Reading the optimistic textbooks and pamphlets of cameralists, it would be hard to disagree with that. Newtonians like Johann von Justi relentlessly promoted the sciences as a tool for transforming the backward lands of the Holy Roman Empire.

But maybe we have taken science too seriously. (For historians of science, it is a distinct possibility.) If you look behind the published literature to the daily business of the empire's many Kammern, it becomes clear that fiscal officials knew how to sell knowledge. Science, like linen or silver, could fill the treasury. In places like Göttingen or Lautern, where there was not much else to offer, knowledge became the biggest regional export product. Professors talked a good game, promising that the latest chemical innovation or financial system could enrich the state. Their livelihoods depended on it. Sometimes, though, they believed their own press clippings, convincing rulers and their ministers to fund ambitious projects. These ventures invariably failed.

This book is not about how knowledge changed the world; it is about how the world changed knowledge. The protagonist of my story is the Kammer, that ravenous fiscal-juridical chamber that devoured everything in its path. History, I am told, is only as good as its sources, and the cameral sciences, which purported to speak publicly about the most secret affairs of the prince, were deeply dishonest. We cannot trust them. And because many of the most important cameral sciences *were* natural sciences, the dishonesty of the Kammer has been inscribed into the literature of science and technology as well. There is no avoiding it.

CHAPTER TWO

Science and Silver for the Kammer

Mining-cameral-science (Berg-Kammeralwissenschaft) is the knowledge, in lands blessed with ore-bearing rock, of how one makes mining prosper, and keeps it prosperous, through wise principles and ordinances, and for the benefit of the state.

—Christoph Traugott Delius, "Abhandlung," in *Anleitung zu der Bergbaukunst.*

Johann von Justi's half brother, Christoph Traugott Delius, wrote the first textbook for Maria Theresa's imperial mining academy in Schemnitz.[1] His textbook included a special treatise on the principles of "mining-cameral-science." The frontispiece captured that concept visually (figure 1). A high-ranking mining official, recognizable by his ceremonial hatchet and cylindrical hat, stands in the foreground. Next to him stands the foreman of the local mine. Miners in the background dig, hammer, and cart raw material, while those in the foreground separate ore from worthless rock. Farther in the background are the smelting huts, whose wafting smoke emphasizes the scene's aura of continual activity. The mining official, simultaneously pointing backward and looking forward, conveys an impression of foresight and omniscience, while the miners, seemingly oblivious to the conversation of the official and their foreman, diligently sort and separate the ore. The foreman, attentive to the directions of the official, remains rooted in the present and the particular, making possible the translation of the official's vision and knowledge into the directed activity of the miners.

FIGURE 1 The well-ordered mine. Frontispiece from Christoph Traugott Delius's *Anleitung zu der Bergbaukunst*, first textbook for cadets at Austria's imperial mining academy in Schemnitz (present-day Slovakia).

Delius's frontispiece portrays an ideally organized eighteenth-century mine.[2] The official, confident in his knowledge and sure of his direction, does nothing and sees everything. The miners do everything and see nothing. Between miner and mining official, between oversight and activity, stands the foreman, the mediator between two worlds and translator of knowledge into activity. Like Justi's "universal cameralist," Delius's mining official existed in a bifurcated world. His was the realm of oversight and direction, while the world of labor and activity belonged to the workers under his direction.

Mining officials were the Ur-cameralists. In Electoral Saxony, Braunschweig-Lüneburg, and Habsburg Austria, where mining formed the backbone of the ordinary revenues, servants of the Kammer directed the mines. Armed with special regalian rights over metals and minerals, these officials managed the great mining regions. It could be dangerous work. When the Kammer stopped providing food and work, miners and metal workers often attacked their superiors. Mining officials responded by encouraging martial discipline in their workers and subalterns. But that did not always work. Everything depended on the *Gewerken,* or investors, who bought shares in the mines and provided regular contributions to keep things running. Without capital from these investors, the mines ceased operating, and wages dried up. Then there was trouble.[3]

The mining towns of Saxony, Hannover, and Austria were strange and remarkable places. Not only were they concentrated centers of wealth, natural scientific knowledge, and big technology, but, as Veit Ludwig von Seckendorff explained, they were zones of dense regulation: "For a peculiar police or community, which must be maintained through discipline and the strict execution of office, arises out of the large mines."[4] Justi, for his part, proposed to make the lowland towns and farms around Hannover, which he found lacking in regulation, more like the mines. "We have barely begun," he wrote in the 1755, "to establish the police institutions and ordinances that can contribute something to the general prosperity; also in this respect various lands appear very empty, so that the trades are left almost entirely to themselves."[5]

Delius's well-ordered mine, fusing good police with the sciences, represented a utopian cameralist vision. If fiscal officials provided the necessary knowledge and direction, the mines would thrive, and the treasury would overflow with silver. That was the promise. Many cameralists served the crown as mining officials and metallurgical consultants.[6] Johann von Justi, the most striking example, served as a mining official (*Bergrath*) in Austria during the early 1750s. Then, in 1765, he entered Prussian service as a *Berghauptmann* (director of mines) for Frederick the Great.[7] Many others came to the great mining districts to promote the "common good" and enrich themselves by serving the Kammer: Leibniz in the Harz Mountains, Goethe in Ilmenau, Abraham Gottlob Werner, Alexander von Humboldt, and Friedrich von Hardenberg (Novalis) in Freiberg.[8] Insofar as each of them shared a vision of the well-ordered mine, they were cameralists too. The knowledge they created fused the fiscal interests of the state with natural knowledge. In their hands, the earth sciences became fiscal sciences.

LABOR, CAPITAL, POLICE

The staggering amounts of gold and silver pillaged and extracted from the Americas between the fifteenth and nineteenth centuries have worked to obscure the memory of Europe's indigenous silver mines. But in much of the Holy Roman Empire, home to most of Europe's major silver mines, mining intensified even as Spanish galleons flooded the continent with countless tons of American gold and silver.[9] During the late Middle Ages, enterprising miners discovered rich silver veins and the occasional gold deposit in regions from the Harz Mountains and the Tirolian Alps to the Erz Mountains and the western Carpathians.[10] But by the late fourteenth century, the more easily accessible silver veins were largely exhausted, and Europe experienced a severe shortage of bullion. By the end of the fifteenth century, as technical improvements allowed mining at greater depth, rich new silver veins were discovered in the same mines that had been abandoned during the twelfth and thirteenth centuries. Deserted mines thus "came back to life" as new equipment and techniques facilitated mining at greater depths than before.[11]

Ironically, as the mines around Freiberg, Clausthal-Zellerfeld, and Schemnitz began producing unprecedented quantities of silver, the Spanish discovered the Bolivian mines around Potosí, which yielded silver in previously unimagined quantities and helped trigger the dramatic inflation of the sixteenth century.[12] Silver now plummeted to less than half of its former value.[13] The inflation and the concomitant drop in the value of precious metals caused many smaller mines to be abandoned, and domestic production suffered. By the beginning of the Thirty Years' War, central Europe's mines were already in sharp decline. By the end of the war, when decades of conflict and depredation had caused most mines to be abandoned, silver production dropped to its lowest point. It would be another hundred and fifty years before the mines again yielded as much silver as they had in the mid-sixteenth century (see appendix 1).

By the seventeenth century, a few great mining districts dominated central European silver production. In Saxony's Erz Mountains, where rich silver veins had been discovered in the twelfth century, Freiberg became the administrative center of a substantial Saxon mining apparatus.[14] In the Harz Mountains, where the Guelfs and their descendants in Electoral Hannover and Braunschweig-Wolfenbüttel exploited an area of rich silver deposits, the neighboring towns of Zellerfeld and Clausthal became the administrative center of Harz mining.[15] To the east, in the Carpathians, the legendary silver and gold deposits of "lower Hungary" helped transform Schemnitz into the administrative and scientific center of silver mining in Habsburg Austria. Even as other metals—iron, copper, tin, lead, zinc—became more important, the legacy of the great silver mines continued to shape the administrative and legal structures of all metal mining.[16]

Mining ventures needed substantial investment to fund the construction of water-pumping operations, drainage tunnels, artificial ponds, smelting works, and other infrastructure. Accordingly, German silver states attempted to lure groups of foreign investors—the Dutch were favorites—to provide money.[17] In order to draw foreign investment, the Kammer promised rich returns. Such promises, however, rested on an idealized picture of the well-ordered mine, complete with trustworthy officials, diligent miners, and rich silver veins. It was not a completely honest picture, but it certainly was effective: by the late seventeenth century, the mining districts around Freiberg (Saxony), Clausthal-Zellerfeld (Hannover and Braunschweig), and Schemnitz (Austrian Habsburg lands) had become concentrated centers of labor, capital, and regulation.[18]

In the well-ordered mine—the paper mine of cameralist reveries—the territorial sovereign and his officials ruled absolutely.[19] As holder of a sovereign "mineral privilege" (*Bergregal*), the ruler leased monopoly rights over metals and minerals to investor groups, and received payment in return.[20] Like the crown lands, the *Bergregal* was property, and it entitled territorial rulers to collect rent on all silver, gold, copper, tin, cobalt, aluminum, and other metals discovered within their borders. This mineral privilege gave princes and their ministers an incentive to promote mining, because the Kammer always got a percentage of the yield. Accordingly, the German silver states offered incentives for mining. Miners, for example, held special privileges from the sovereign, forming a class of subjects free from prosecution by the civil courts. In return for their special rights and privileges, they were supposed to submit to the harsh conditions of their work and the direction of the state's mining officials. The "principle of direction" (*Direktionsprinzip*) authorized fiscal officials to promote and control all aspects of metal mining and smelting.[21] If the prince hoped to exploit his mineral possessions fully, his officials had to know what was being extracted from the territory's mines and how it was being processed. They also had to provide for the necessary infrastructure—great drainage tunnels, holding ponds for water power, legal guarantees, grain depots—that would to bring miners to the mountains and attract foreign investment.

Mining officials were not merely "technical" personnel; they were the governors of the mines, and they aimed to rule them absolutely. During the eighteenth century, these ambitions were mirrored by increasingly elaborate parade dress uniforms. Saxony's general commissioner of mines (*Generalbergkommissar*), Friedrich Anton von Heynitz, the apotheosis of this "quasi-military" type, recognized the affinity between mining officials and military officers.[22] The direction of a mining office, he explained in a letter to the elector, demanded "almost martial subordination and complete confidence in its chief."[23] For Heynitz, the success of a large mining district depended on the martial bearing of its officers.[24] Figure 2 shows Heynitz in full parade dress. In his right hand is a *Bergbarte* (ceremonial hatchet) engraved with pictures representing every aspect of the "mining household."[25] The left hand rests near his sword. The epaulets and golden galloons signify his rank, and his hat bears the representative marks of Saxony's elector: black and yellow cockade (the dynastic colors) and coat of arms.

> His hair, or wig, is cut *à l'abbé,* just like all of the mining and smelting personnel, and covered with a round green satin cap, which has the form of a regular hat and is bordered with broad stripes. The elector's coat of arms is stitched upon two shields on the front of this cap, and on the left side is a yellow and black cockade with a white and black plume above it. The jacket has a scarlet red collar, lapels, and borders which, like the button holes, pockets and seams, are stitched with broad golden galloons. Over and under the folds of the arms are epaulets of golden fringe, complete with *Crepins* and *Bouillons.* The miner's pouch, miner's leather, and knee-pads are made of Cordovan leather and stitched with golden lace.[26]

Like the police ordinances that tried to dictate every aspect of behavior in and around the mines, the careful delineation of white taffeta stockings and black silk tassels aimed to direct conduct and to link center to periphery. The symbolic trappings of the miner—leather knee-pads, miner's leather, miner's pouch—linked the official to the local sphere of the mines. At the same time, the symbols of sovereignty, like the coat of arms and the cockade, tied him to the central sphere of court and Kammer. Miners, on the other hand, bore few of the sartorial privileges of officialdom—no lace on the shirt, no sabre, no royal coat of arms, no plume. "The common miner is, in respect to his clothes, completely different from the parade dress of the officials. His cap is of cloth without any decoration, except for a small round yellow and black cockade in front over the forehead. His jacket is the well-known miner's smock, which he normally wears, and has to wear."[27] Like the paper universe of police ordinances and cameralist texts, parade dress uniforms suggested a world of omniscient officials and obedient miners. Today, the traditional miners' parades in Freiberg or Banská Štiavnica, now staged mainly for German tourists, seem to carry echoes of a disciplined past. It is easy to forget that these parades were also staged three hundred years ago, but for a different audience.

FIGURE 2 Portrait of Friedrich Anton von Heynitz. Printed with permission of the TU Bergakademie Freiberg.

Between parades, when dukes and princes had left for home, there was plenty of theft and corruption. Large quantities of silver went missing every year, and the Kammer responded by promulgating a dense web of police ordinances.[28] It is unclear whether these ordinances actually accomplished much; renegade smelters and assayers, corrupt foremen, and dishonest contribution collectors did not go away.[29] It is clear, however, that efforts to police behavior spawned a paper world of order and honesty, full of obedient miners and faithful officials. Seckendorff's model principality, for example, was home to an ideal mine, complete with good cameralists. These officials, selfless caretakers of the prince's mineral privilege, fused the knowledge of nature with a passion for order. This imagined world contrasted sharply with the real situation of the mines, where officials took bribes and, as Seckendorff himself acknowledged, common miners were given to "cursing, boozing, carousing and brawling."[30] Cameralists like Seckendorff liked to imagine a better world in which dutiful officials, armed with the knowledge of nature and the will to discipline, ruled over obedient miners while placing the "common welfare" before their own interests.[31]

THE MINE AS MODEL STATE

When mining officials turned their attention to the state, they tended to see the well-ordered mine writ large. They sought to organize artisans, farmers, burghers, and merchants much as a Berghauptmann might direct the miners and investors in his district. They saw in laws and police ordinances the valves and pistons of the great machines that dominated the landscape of the silver mining districts. They believed that every universal cameralist ought to catalogue the territory and its resources, just as every Berghauptmann was expected to understand all parts of the mining operations under his direction.

When Heynitz turned his attention to political economy in 1775, for example, he treated the state like an oversized mine.[32] His *Tabellen über die Staatswirthschaft,* which appeared anonymously in both French and German, distilled Saxony and its administration into a series of tables—tables on population and employment, agriculture, income and expenditures, and the inflow and outflow of specie. "If princes could survey the whole state œconomy of their lands through such tables," wrote Heynitz, "I believe they would fulfill their duties better." State administration was simply another form of mining administration. "The man entrusted with the œconomy of the state cannot confine himself to mere theory. The details of all its parts, the successes and failures of previous administrations, the examples offered by the exact study of similar state administrations; these are the things that, more than any others, must educate and instruct him."[33] Heynitz's young "political economist," that is, would be trained like a mining official, learning about each part of the mining operation in detail.

Like Heynitz, Justi had the instincts of a mining official. He likened the state to a machine and the *Nahrungsstand*—a term meant, in this context, to denote society's laboring classes—to a great engine.[34]

> The *Nahrungsstand* is the driving engine of the great machine of the state. Each occupation must assume its proper place therein and contribute as much to the movement of the whole as the well-being of the commonwealth demands. Each type of occupation (*Nahrungsart*) must constantly serve to support and further the others, and none may be det-

rimental or burdensome to the others. In order to maintain this crucial coherence of the *Nahrungsstand,* many laws and regulations are necessary; and these laws are the object of police.[35]

Justi's "machine of state" was modeled on the elaborate water-powered machines he had encountered in the Habsburg mines outside of Vienna, technical marvels of the German and Austrian lands. The cameralist, at once mechanic and police official, had to design and engineer the ideal *Nahrungsstand* by constructing the social machine best able to harness the collective energy of its members. Good police would arrange and direct the mass of society—its merchants, peasants, burghers, craftsmen, manufacturers—in the most productive possible way. It was an ambitious dream.

Justi divided the state, like the mines, into two great realms: directors and directed, learned and ignorant, *Gelehrtenstand* and *Nahrungsstand.* The *Gelehrtenstand,* learned professionals and state officials, the keepers of knowledge and lords of direction, existed on one side of the great divide. The countless members of the *Nahrungsstand,* that great mass of productive activity, existed on the other. Neither realm could subsist alone. The health of the *Nahrungsstand* depended on "the industriousness of the people" on the one hand and the "rational direction and leadership of all *Nahrungsarten* and occupations" on the other.[36] The "directive class" of *Gelehrten* and state officials would remain sterile without the *Nahrungsstand,* and the *Nahrungsstand* would remain a chaotic mess of productive activity without proper direction and control. A *Nahrungsstand* without direction was blind; a *Gelehrtenstand* without productive activity was empty.

SCIENCE AND THE SILVER STATE

The well-ordered mine may have looked good in textbooks and police ordinances, but the world of everyday administration was something altogether different. No one could be trusted. In Saxony, the archetypal German silver state, reformers spent the better part of two centuries trying to fashion a transparent mining administration that would serve the fiscal interests of the elector. They failed. Heynitz, the most forceful of these reformers, finally founded the Bergakademie Freiberg, the first mining academy in the world, as a way to create the right kind of mining official. Historians typically treat Freiberg as a training ground for "German mining engineers," technical experts who used the latest science and technology to improve yields and efficiency.[37] That is misleading. Heynitz's young "cadets," the students at the mining academy, may have learned some useful things about water wheels and subterranean geometry, but that is not why Heynitz founded the place. He established the mining academy because he needed someone to trust.

Heynitz had been brought to Saxony to help fix the state finances, which had collapsed completely in the wake of the Seven Years' War. Many blamed Count Heinrich von Brühl for Saxony's fall. The French ambassador in Dresden even joked that Brühl's administration had done as much damage to Saxony as the King of Prussia.[38] Justi's satirical biography of Brühl, which appeared anonymously in 1760, painted the man as a self-interested, unsystematic, haphazard creature of the court. For Justi, as we have seen, Brühl was the quintessential bad cameralist.[39]

By the end of the Seven Years' War the heir to Saxony's throne, Prince Friedrich Christian, and his circle of advisors in Dresden had begun to conceive a wholesale reform of Saxony's fiscal-administrative system. In November 1761, the leader of the group, Thomas von Fritsch, began sending memoranda to Count Brühl in Warsaw.[40] By the spring of 1762, Brühl and Fritsch were corresponding about the establishment of a special commission that would "consider everything precisely and submit projects for His Majesty's approval about how to assist this impoverished land and its subjects in every possible way."[41] By April, a Restoration Commission chaired by Fritsch and constituted of other officials, set to work. One of the first issues to be addressed by Fritsch and his commissioners was the improvement of Saxony's mines and mining administration.

In March of 1762, while negotiating over the nature and composition of the Restoration Commission, Fritsch sent an urgent memorandum to Warsaw about Saxony's mines. "Mining," wrote Fritsch, "is undeniably one of the most important, if not the single most important, pillar of this land's welfare; its repair and maintenance, therefore, deserve the most exact reflection and the most careful consideration." Fritsch urged the Kammer to prepare a comprehensive table of mining income and expenditure. Such an overview, he argued, would demonstrate "how important mining is for the land, and how necessary it is to keep a diligent and watchful eye on the same." Fritsch complained that the mines had suffered from bad administration. Foreign investors had lost faith in Saxony's mines. Trust had to be reestablished through a mining administration marked by the "strict oversight of the sovereign." "We lose this trust," he explained, "if we appoint bad or dishonest officials."[42]

Considerations like these prompted Elector Friedrich Christian to add a powerful new voice to Dresden's Kammer. At the end of 1763, he appointed Heynitz, by then an experienced senior mining official from

Braunschweig-Wolfenbüttel, as fourth mining councillor in the Kammer.[43] Heynitz took up his new post in February of 1764. The "*Kammer- und Berggemach,*" as it was called, explicitly joined Saxony's financial administration to its mining administration; it was the most powerful fiscal collegium in Saxony. Heynitz, who had been extremely frustrated by his lack of power in Braunschweig-Wolfenbüttel, hoped that the new position in Saxony would give him more autonomy and authority.

The week after Heynitz's appointment, Elector Friedrich Christian died suddenly, leaving his brother, Prince Xaver, as regent until the young heir, Friedrich August, came of age to rule. Xaver, however, soon fell out with Fritsch and his allies. Heynitz, meanwhile, encountered resistance from his colleagues in the Kammer almost immediately upon his arrival in Dresden.[44] In search of greater influence, he sent Cabinet Minister von Einsiedel a memorandum on the proposed reorganization of Saxony's mining administration.[45] The plan succeeded, and Heynitz was appointed "general commissioner of mines" (*Generalbergkommissar*) in June of 1765.

Though the new position did not give Heynitz complete control over Saxony's mining administration—many of his plans and projects remained subject to the approval of the Kammer in Dresden—it did give him considerable influence in Freiberg, where he became head of the Oberbergamt (Chief Mining Office), together with Friedrich Wilhelm von Oppel.[46] Still animated by the ambitions that had originally brought him to Saxony, Heynitz turned his attention to the Oberbergamt. It was at about this time, in the summer of 1765, that he seems to have embarked on a new approach. If he could not shape Saxony's mining policy from above, in Dresden, then he would reform it from within by taking control of the regional appointment and education of Saxony's mining officials. He would, that is, create a generation of officials in his own image.

In September 1765 Heynitz sent another memorandum to Einsiedel, expressing concern about the poor condition of the Oberbergamt.[47] More particularly, he discussed the poor quality of the mining officials who worked there, and the "lack both of those who are now usefully employed, and of those who can be recruited to direct affairs." Heynitz felt that the situation in Freiberg was chaotic and unacceptable. Since no one had a view of the whole, Saxony's mines were in complete disarray. He proposed to remedy the situation through wholesale reorganization of the mining administration which, he argued, should be arranged according to the four natural divisions of the mining œconomy: mining proper, stamping and separation, smelting and assaying, and record keeping and acquisition of materials. Each "talented subject" could devote himself to one or another branch of mining operations.[48] Heynitz urged Einsiedel to issue an "instruction" to the Kammer about the reorganization. The new arrangement, he argued, would help to curb abuses and encourage industriousness, allowing for more effective oversight since each official would be responsible for a discrete aspect of the mine.[49]

Heynitz then turned to a specific enumeration and critique of the mining officials in Saxon service. Berghauptmann von Ponikau, at sixty-three, had "little life left in him." Mining Councilor von Wiehmannshausen was not only old, at almost sixty, but had been hampered by a "gouty foot" for many years. Commissions-Rath Meybach was no better. He was also about sixty, and, with a only a smattering of knowledge in "speculative chemistry and hydraulics," was quite worthless for the practical work of managing a fiscal collegium. Mining Councilor Pabst von Ohain had more promise.[50] He had the requisite knowledge, insight, vigor, and zeal. Unfortunately, complained Heynitz, Ohain did not "seem wholly free of the passions, shows too much politics, never follows the truly straight path, doesn't allow himself to be led, and shows even less evidence of being able to lead others." All of this led Heynitz to the conclusion that there was a "real shortage of capable people to fill posts as mining officials and mining councillors in Freiberg."[51] Heynitz then proposed a solution to the problem:

> The mining district in Freiberg has established a scholarship fund, from which sons of the state's mining officials [*Bergbediente*] can get money to learn subterranean surveying and assaying as training for subaltern positions. This fund is of great use, and I have already proposed many times in the *Berggemach* . . . that people who apply themselves to mechanics and other similar sciences should get something from this fund. But because this fund is not adequate for the *kind of people* who want to learn the mining sciences *in order to direct the operation* (for which reason there are more considerable funds in Hungary, Austria, Bohemia, Sweden and in the Harz), and without [adequate funds] it is not easy to guarantee that anyone will take on this always costly profession, I see it as my duty to point out that His Royal Highness might see fit to increase the fund with a contribution from the treasury.[52]

Heynitz was already preparing the way for a mining academy. The existing scholarship fund, which had been in place since 1702, no longer seemed adequate to him. It had been designed to provide narrow technical training for subaltern officials, that is, training for the wrong type of official. His proposal, on the other hand, sought to provide funds for educating a completely different kind of mining official. It aimed, that is, at cultivating cameralists to direct the mines.

Heynitz proposed a three-year period of training. The first two years would be spent in one of Saxony's mining towns, probably Freiberg, and candidates would receive 200 taler annually. During this time candidates would study under the direct supervision of the *Oberberghauptmann.* In the third year, the scholarship would increase to 400 taler, and the candidate would begin touring mines outside Saxony. In certain respects, the proposal resembled the structure of the 1702 scholarship fund, which had provided state support for aspiring young assayers and subterranean geometers to learn a trade from skilled subaltern officials. But Heynitz's plan was significantly different, for it aimed at cultivating high-level officials for service in central bureaus like the Oberbergamt and the Kammer.

Heynitz not only planned to train a new generation of mining officials but also aimed to weed out "useless" officials from Saxon service. He had, for that purpose, begun to prepare an overview, in tabular form, of salaries and other income for all of Saxony's mining officials. He also suggested that "many official posts [*Bedienungen*] can be eliminated or combined." In other cases it was simply a matter of dumping old, tired, corrupt, and useless officials for a new generation of better ones.[53] The success of the mines, he believed, depended on cultivating the right kind of official. In fact, Heynitz later claimed that the health of Saxony's mines rested completely on God's blessing and on the "diligence, insight, seriousness, application, liveliness and integrity of the land's mining and smelting officials."[54] He attributed the decline of mines in the Harz and the Hungarian Carpathians, for example, to the absence of good officials there. A small investment in the education of Saxony's mining officials, he felt, would directly benefit the treasury. "This proposal," he promised, "will soon yield a rich profit, and your Excellency is already personally acquainted with the importance of Electoral Saxony's mines."

Two months later, Prince Xaver and the elector's widow, Maria Antonia, visited Freiberg. Heynitz, hoping they would support his plans for a mining academy, put on a show.[55] He had the mine shafts illuminated and had the miners' tools restored. He arranged demonstrations of ore stamping and separation. He ordered two officials, Christlieb Ehregott Gellert and Friedrich Wilhelm Charpentier, to perform chemical experiments.[56] And he organized a dramatic miners' parade to follow the evening meal. Xaver, who had a weakness for military processions, authorized Heynitz to write up a concrete proposal for the mining academy on the spot.[57]

Heynitz's plan for the new academy did not merely extend the purposes of the existing scholarship fund. Rather, the new mining academy, as he envisioned it, would prepare young members of the nobility for careers in the upper echelons of Saxony's fiscal administration.[58] Whereas the existing scholarship fund had been established to support the acquisition of technical skills, especially assaying and subterranean surveying, the new academy would provide broader training in natural history and natural philosophy. The cadets were also expected to have legal training, and the plan provided for some university study in jurisprudence, financed by the sovereign. Moreover, Heynitz strongly believed in the value of touring the mines, and his plan thus set aside almost half of the total budget for travel costs. His new academy would educate young officials for service in Saxony's fiscal-mining bureaus—especially the Kammer and Oberbergamt—well into the future. Provided with noble titles, legal training, and well-placed connections from their extensive travel, the cadets were being groomed for positions in the upper levels of Saxony's fiscal administration. Unlike their predecessors, who had used the scholarship fund simply to learn specific skills, the Bergakademie Freiberg had more ambitious goals. It would produce good cameralists to direct Saxony's fiscal-mining state.

Like Justi, Heynitz promised a third way between the empty theory of the universities and the artisanlike practice of the Kammer. He disdained the "Medicos and Chymicos" in Dresden and their useless, speculative sciences.[59] But he was equally critical of officials like Count Brühl who, through family connections and sycophancy, had risen through the administration to positions of power and influence. Heynitz institutionalized this vision in the new mining academy. Older forms of "artisanal" training for assayers gave way to a more thorough grounding in the natural history of minerals. Heynitz encouraged the systematic chemical analysis of minerals, while others, such as Christlieb Ehregott Gellert, a smelting official and early teacher at the academy, turned to more general questions about the behavior of metals in fire.[60] At the same time, the mining academy was insulated against charges of empty theoretical speculation. Its location in Freiberg, at the heart of Saxony's mining region, offered exposure to the everyday life of the mines. Those who went to study there, often from large cities or university towns, experienced not only the technical wonders of Freiberg—its drainage tunnels, stamping works, smelting huts, and water-pumps—but also the "peculiar police" that characterized every mining district. In addition, the academy's teachers and administrators were themselves mining officials, and their participation in the regular sessions of the Chief Mining Office (*Oberbergamt*) and the Chief

Smelting Office (*Oberhüttenamt*) linked them to the administrative life of the mines.

The Bergakademie Freiberg, as Heynitz had conceived it, posed a challenge to the universities. The task of educating officials for state service, whether in law, medicine, or theology, had traditionally been the exclusive province of university education. With its new mandate, the mining academy now began to train its own officials.[61] Moreover, Freiberg's increasingly systematic instruction in mineralogy and chemistry offered an alternative to university education, which typically treated these subjects as auxiliary sciences for the medical faculty.[62] But the liberation of such sciences from the medical faculty signaled at the same time their subordination to Saxony's Kammer- und Berggemach. The mining academy, that is, gave the Kammer direct control over certain kinds of knowledge and bypassed the troublesome old universities, with their special academic privileges and quasi-autonomous faculties.

Heynitz's correspondence indicates that he wanted Freiberg's professors to be directly accountable to the Kammer and its fiscal officials. In January of 1769, for example, he wrote to the elector, requesting that he and Oberberghauptmann Oppel be named "curators of the Freiberg *Bergakademie* with absolute power." Heynitz understood this as the power "to pay teachers according to their acknowledged industriousness, and in cases of flagging industriousness to reduce that salary" and to give the students more or less financial assistance according to their diligence and merits.[63] He intended to divest professors and students of all autonomy. The reform-minded Saxon ministry, for example, considered the new Bergakademie a welcome alternative to the troublesome University of Leipzig where, as the head of the Restoration Commission, Thomas von Fritsch, put it, "the university spirit of strife [*Universitätszanckgeist*] is already too much in vogue and has increased hindrances to the police there."[64] Everything in Freiberg would depend on the will of the curator, who was himself a fiscal official. It was the kind of control that even the most powerful university curators could only dream about. Students and faculty at the new mining academy were subsumed directly into the structure of state administration, and Heynitz and Oppel, the heads of the Oberbergamt, assumed unchallenged control over its daily administration.

The subsumption of the Bergakademie into the state's fiscal apparatus achieved its fullest visual display in Saxony's periodic mining parades.[65] In the parades, the top-ranking mining official rode in front, followed by columns of the various groups of miners and smelters. After 1766, science, represented by the students and professors of the mining academy, joined the parade, marching in step with other members of the mining administration. The cadets, as their parade dress uniforms suggested, were high-level mining administrators in training.

Heynitz and Oppel used many of the same incentive structures to control and motivate cadets as they did in directing the mines. They developed, for example, a system of academic piecework incentives. Cadets would receive monetary rewards for good work—models, drawings, experiments—done in residence at the Freiberg academy.[66] Officials trained in the principles of police science, natural history, chemistry, technology, and natural philosophy would reinvigorate Saxony's mines through improved direction and oversight. A new generation of officials would fill the sovereign treasury with silver from the local mines and capital from Dutch investors.

The Bergakademie fulfilled a vision that cameralists had been articulating for decades.[67] Systematically trained officials, they argued, would render the mines more responsive to the Kammer. Imbued with good principles, they would make the mines more efficient, more open, and more honest.[68] Carl Friedrich Zimmerman, for example, promised that cameral science would confer "lasting order" on mining, so that every "high Kammer collegium" would "establish better order in its administrative bodies and avoid losses and fraud."[69] Well-ordered knowledge would beget profitable mines. Zimmerman's *Obersächsische Berg-Academie* described the curriculum of a proposed mining academy, whose purpose he made quite clear: an "increase in sovereign revenue, and a blessed profit for those *Gewerken* that invest actively."[70] Justi too articulated the importance of mining academies during the 1750s, a decade before Freiberg was established. "That the mining sciences prosper," he wrote, "is not unimportant, and one must therefore provide good instruction in both universities and in special mining academies."[71] But the direct inspiration for establishing a mining academy in Freiberg probably came from Heynitz's friend, the cameralist Daniel Gottfried Schreber, a longtime proponent of such institutions.[72]

Freiberg may have been the first of its kind, but it was quickly followed by imitators in Berlin and Schemnitz in 1770.[73] These mining academies in Berlin and Schemnitz patterned themselves on Freiberg. (Prussia even dispatched a spy to report on the new mining academy.[74]) Like the Bergakademie Freiberg, these academies sought to train cameralists. Archival evidence in Berlin and Wroĉlaw (Breslau) suggests that the impetus for founding them arose largely from a dissatisfaction with the universities; mining academies offered a way to sidestep the universities by making certain kinds of knowledge and training directly subject to the state's fiscal bureaus.[75]

Prussia, especially, was in need of skilled mining officials after the Seven Years' War. The acquisition of mineral-rich Silesia in 1741 had created an administrative vacuum in the mining districts there, and Berlin was still playing catch-up some three decades later. That is why Frederick the Great commissioned Johann von Justi as a Prussian Berghauptmann in 1765.[76] Always attuned to the administrative innovations of neighboring silver states, Saxony and Hannover, Frederick's ministers took great interest in the new Bergakademie Freiberg. Initially, they responded by trying to reform the universities. In 1768, for example, professors at Prussia's four universities—Halle, Frankfurt an der Oder, Königsberg, and Duisburg—received direct orders from Berlin to offer lectures in the mining sciences.[77] Later, in 1770, Ludwig von Hagen, then the most powerful of Frederick's officials, spearheaded a new initiative dedicated to the mining sciences. Prussia, he argued, desperately needed to improve the quality of its officials, because of the "general uselessness of so many state servants." The situation, he continued, was especially bad in the "whole œconomic field," since "so many idiots and ignoramuses in these and related subjects . . . continue to be supported and salaried." According to Hagen, these "idiots and ignoramuses" lacked the kind of knowledge required of every good œconomist or mining official: "physics—especially the practical part—mineralogy and metallurgy, applied mathematics, forestry."[78]

In the following months, Prussia's mining and smelting department became directly involved in university affairs. Minister von Fürst directed the consistory to implement teaching of the mining sciences. These officials went so far as to collect syllabi from Halle, Königsberg, and Frankfurt an der Oder. Surviving copies of these syllabi are filled with the red-pencil underlining and the marginal notes of Prussian officials.[79] These red-penciled syllabi testify to the direct influence wielded over the sciences by Prussia's fiscal-police state.

Not to be outdone, Maria Theresa's government established its own mining academy in Schemnitz. Here, as in Berlin and Freiberg, the purpose was to cultivate good cameralists.[80] Delius's textbook, with its treatise on *Berg-Kammeralwissenschaft,* reflected the larger purposes of the institution: "Mining-cameral-science is the knowledge, in lands blessed with ore-bearing rock, of how one makes mining prosper, and keeps it prosperous, through wise principles and ordinances, and for the benefit of the state."[81] Delius, one of the first teachers at Maria Theresa's new mining academy, wrote his textbook to fulfill his official orders from Vienna. In the Austrian lands, as elsewhere in central Europe, the systematic education provided by the new mining academy in Schemnitz represented a significant change. For hundreds of years, Austria's mining officials had been trained directly in the state's Kammern. They had learned on the job or, as Delius put it, "through long experience without principles." He considered this lack of systematic education responsible for the fact that Austria's mines were "not directed with appropriate insight, order and œconomy." Now Austria's mining officials would be trained systematically in all "theoretical as well as practical mining sciences." The advantages of the new academy, Delius promised, would spread to all the mines in the land as a new breed of mining official, one blessed at once with "principles and experience," began to populate Habsburg Austria's mining districts. Like his half-brother, Justi, Delius sought to train professional cameralists. Practical experience alone no longer sufficed for those who had to organize, arrange, and police the mines. Delius's ideal mining official, armed with systematic principles and a moral calling to promote the general welfare, was another version of Justi's good cameralist.[82]

Delius singled out a fiscal official, Minister Franz von Noworadsky Kollowrat, president of the *Hofkammer,* for special praise in the preface. Kollowrat, he explained, was a model official who understood "mining, minting, and fiscal matters" as well as the œconomic and technical features of the mines. It was Kollowrat who had "drafted a plan for the establishment of a proper mining academy where young people . . . could receive a thorough practical and theoretical education in all parts of the mining sciences. Our most wise monarch, who always supports useful proposals, blessed the plan with her approval." The academy in Schemnitz, organized much like the Bergakademie Freiberg, had been established for the same reasons.[83]

Unlike Austria, Prussia, and Saxony, Hannover did not establish a mining academy of its own.[84] Minister Gerlach Adolf von Münchhausen, curator of Hannover's modern university in Göttingen, had different ideas. He had worked for decades to establish a cameralist faculty in Göttingen, complete with auxiliary sciences, such as agriculture and subterranean geometry.[85] Cameralists like Justi and Johann Beckmann, who received strong institutional support from Curator Münchhausen, offered lectures and wrote books on the mining sciences.[86] Other Göttingen professors, like Abraham Gotthelf Kästner and Johann Friedrich Gmelin, focused on topics related to mining.[87] The holdings of Göttingen's university library demonstrate that Münchhausen and his administrative successors were dedicated to promoting the mining sciences, for the library possesses a first-edition volume of virtually every important seventeenth- and eighteenth-century text on mines and mining, among them the major works of Freiberg's professors. Freiberg's first textbook, *Bericht vom Bergbau,* came to Göttingen as a gift from Heynitz.[88]

The university acquired others systematically through local booksellers like Dieterich and Vandenhoeck. In general, the collection resulted from carefully targeted acquisitions by Münchhausen and the university's librarian, Christian Gottlieb Heyne (see appendix 2).[89]

German silver states, then, moved to train cameralist mining officials in the aftermath of the Seven Years' War, whether in universities or specially-designed mining academies. Historians have generally treated these academies as "utilitarian," the assumption being that such places produced, and were intended to produce, skilled engineers and metallurgists, experts who could apply science to technical problems. Nothing could be farther from the truth. The first mining academies, all modeled on Freiberg's, were intended to produce cameralists, that is, officials who could direct and oversee the great œconomy of the mines. The fiscal-natural knowledge produced in mining academies and fashionable universities (like Göttingen's) served these same purposes. In Saxony, for example, Heynitz mobilized the mining academy and its cadets in a larger effort to reform the electorate's fiscal administration.

HEYNITZ'S AUDIT COMMISSION, 1766–1771

Prince Regent Xaver placed Heynitz at the head of a special audit commission in November of 1766, soon after the Bergakademie Freiberg opened its doors. The other members of his commission included Oberberghauptmann von Oppel, Berghauptmann Carl Eugen Pabst von Ohain, Mining Councilor Johann Polycarb Leyser, and Friedrich Wilhelm von Trebra, a young cadet from the first class of the Bergakademie. Heynitz and his commission received orders to tour Saxony's mining districts and to draft suggestions for improving them. The inspection tour took the commission two years to complete. It took more than two additional years for Heynitz to submit the final report.[90] After establishing the Bergakademie to cultivate a generation of officials in his own image, he wanted to fill Saxony's fiscal and mining collegia with cadets from the mining academy. As head of the audit commission, he was finally in a position to effect such change.[91]

The commission began its work by issuing a "conspectus," or comprehensive survey, to all of Saxony's mining districts. Officials from the local districts were required to make point-by-point responses to a battery of questions, to prepare tabular overviews of yields and expenditures, and to provide lists of personnel. The commission's orders also required local officials to submit suggestions for improving the mines in their districts. After this first stage was complete, the commission began formally to examine and verify the information. The auditing process took place on-site and in the presence of those local officials who had prepared the survey responses.[92]

Though the commission made substantive proposals about material issues, much of its work was dedicated to discovering and eliminating bad officials, the presumptive source and cause of bad mines. Heynitz's commission was also a kind of fiscal inquisition: if threats and exhortations could not change behavior, the commission would request "dismissal from service" and suggest a replacement (128). Such replacements, of course, would come from the new Bergakademie. Heynitz's final report supported these changes, explaining that the general quality of Saxony's mining officials had long suffered from the lack of attention devoted to the "cultivation of the auxiliary sciences" (130). In the universities, preparatory courses for students in the higher faculties often consisted of "auxiliary sciences." Students in the medical faculty, for example, would study chemistry and botany as part of their professional training. Heynitz's Bergakademie resembled a free-standing cameralist faculty, a professional school with a specially tailored structure of auxiliary sciences. As with law or medicine, however, these sciences—the cameral sciences—served to promote a particular professional identity. For mining officials, that identity depended on the ability to generate profit through solid œconomy and good police. It aimed to produce officials who could generate "the greatest profit with the most consistency in the shortest time through the simplest means." But Heynitz and his colleagues recognized that written ordinances and administrative directives alone could not guarantee success. One needed "constant interlocking controls" and mechanisms of oversight (133).

These interconnected mechanisms of oversight formed the basis of the "peculiar police" that aimed to shape life in the mines. More often than not, however, visions of good police failed to generate well-ordered mines, and local officials took the blame. In the small districts of Ehrensfriedensdorff and Geyer, for example, the commission noted that tin and silver yields had dropped precipitously. Its final report blamed the local officials for that decline: "The quarrelsome self-interestedness and spirit of denunciation that spread among the mining officials had, through the disunity and mutual hostility it created, hindered the implementation of good measures; so that one would almost have forgotten there was mining here, were it not for the large heaps of waste rock" (144). The commission discovered similar problems elsewhere. In the important district of Schneeberg, home to valuable cobalt deposits, it discovered "self-interested" officials. Moreover, the *Bergmeister, Bergschreiber* and *Zehender* there had constantly registered complaints against one another. When the *Bergmeister* failed to comply with orders from the authorities in Freiberg, he was dismissed (148–49). In Eibenstock, meanwhile, a general indolence prevailed: the foremen were ignorant, there were no drawings of the mines, and, perhaps worst of all, there was "no index in the archive" (151).

Heynitz intervened personally to install cadets from Freiberg in important official positions. Friedrich von Trebra, the Bergakademie's first student and a member of the Audit Commission, was one of them. Trebra had been in Freiberg a little over a year when he was called upon to direct the mines in and around Marienberg, an important mining district in the Erz Mountains. Only twenty-eight years old at the time, he was surprised by the appointment. "That I might be the person for this position did not even enter my thoughts, and I was extremely surprised when I received an order early one morning to meet Generalbergkommissar von Heynitz." At the meeting, Heynitz and Oppel asked Trebra whether he would accept the position of *Bergmeister,* essentially district chief, in Marienberg. Trebra replied with a question, asking Oppel and Heynitz whether they really thought he was ready to serve as *Bergmeister.* Heynitz explained that Trebra's reputation as "diligent and honest" was most important "since the mining officials, and with them the mines, had lost their good reputation with the public due to assorted misdeeds." Moreover, the Marienberg district had fallen so far that there was not much to spoil. If Trebra proved "diligent and industrious" and revived the condition and productivity of the mines, he could expect rewards and advancement in the near future.[93]

Heynitz appointed the young cadet because Saxony's existing officials lacked "knowledge, application, and integrity." Trebra was also a nobleman (18). The affair sheds light on the true purposes of the Bergakademie, which had been designed to do much more than simply transmit technical knowledge. Students, carefully chosen, would learn not only about chemical principles and geology, but also how to coax more work out of recalcitrant miners and entice investors.

THE DISORDERED MINE

What Trebra encountered in Marienberg was nothing like the well-ordered *Bergstaat* of cameralist reveries. It was dirty, dangerous, and completely chaotic. When he arrived in Marienberg, he discovered a ghost town. The streets, lined with burned-out houses, were empty. The mining administration there was even worse. Its account books, registers, and inventories were not only indecipherable, but also incomplete. The previous *Bergmeister* had taken many of the records with him, and the remaining officials had ceased to distinguish between their private affairs and official business. The shift foremen, for example, had taken to paying wages in their own houses instead of at the mining office. For Trebra, this confusion between public and private was the worst thing about Marienberg, and the surest sign that things were in disarray. Trebra recalled the shock of it all. The "disorder" he encountered in Marienberg far surpassed anything he could have imagined (27–77).

Trebra did his best. He tried to introduce order and discipline into the daily routines of the mining office, and he tried to establish accountability among the officials there. At least that is how he recalled it many years later. Clearly, though, the most important thing was reputation. For small German mining districts like Marienberg, reputation was everything, because outside investment depended on it. Trebra recognized that other cadets in his class had been more technically gifted and more experienced, but Heynitz had chosen him. Why? Apparently, Heynitz felt that Trebra, with his family connections and noble rank, would be most successful at raising money from foreign investors. Heynitz chose well. In 1770, Trebra was sent to Amsterdam on a marketing mission. He arrived in Holland with the names of some contacts and some charts that had been translated into Dutch. Amazingly, he was able to convince some of the Amsterdam merchants to invest in Marienberg (12, 199–217). The infusion of Dutch capital was enough to trigger a small renaissance in Marienberg. But it was not meant to last. After years of struggling to explain Saxony's *Bergstaat* to the Dutch, he discovered "how difficult it was to attach those republican merchants to our mining and our institutions" (532).

Three decades later another famous Freiberg professor, Abraham Gottlob Werner, was still justifying the Bergakademie (and its budget) to Dresden's authorities.[94] Many important students, including the Prussian ministers Baron Stein and Count Reden, had heard lectures in Freiberg. Mining officials from across Europe and the Americas had studied there. But that was not the main thing. The mining academy, he argued, drew money to Freiberg by attracting wealthy foreigners. Werner, good cameralist that he was, understood that Saxony's silver came not only from the mines of the Erz Mountains, but also from the eager hands of wealthy foreign students.

Professors in Freiberg were open about the need to attract wealthy students from at home and abroad. It is not at all clear, for example, that Werner's classification of minerals or Novalis's romantic science did much to fill Saxony's coffers with local silver. It is clear, however, that they built a towering reputation for the place and that it was teeming with rich students from all over

Europe by the end of the eighteenth century. Moreover, the Bergakademie managed to keep some of Saxony's own elite, and thus its native silver, close to home. When the Bergakademie opened in 1765, the once legendary yields from Saxony's silver mines had long since been overshadowed by gold and silver from the Americas. But the relatively poor veins of the Erz Mountains did not present an insurmountable problem, because cameralists like Heynitz and Justi had meanwhile discovered how to transmute academic knowledge into silver coin.

We are accustomed to read stories like these in terms of progress; science and discipline remake the human and natural worlds to serve the fiscal demands of a modernizing state. But between the lines lurks a completely different narrative of disorder and desperation, of a Kammer that could not feed its miners or gather enough capital to keep the mines running. In that story, science plays an entirely different role than the usual one: not the engine of modernization, but the ideology of a desperate fisc; not a useful tool for technical experts, but a noble lie spun for Dutch investors.

CHAPTER THREE

The Knowledge Factory

It is true that the greatest and the smallest things—potato growing and the oak tree, the art of preparing manure and the analysis of the infinite, pig breeding and pedagogy—have generally more or less relationship with the art of governing. But does it follow that one must know all of them intimately? And how could that possibly happen? —Anonymous, "Schreiben des Kantors."

"You, Mister Curator, are the factory director; the instructors at universities are the apprentices; the young people who attend them, and their parents and guardians, are the customers; the sciences taught at those universities are the wares; your king is the lord and owner of his scientific factories [*wissenschaftliche Fabricken*]."[1] Friedrich Phillip Carl Böll, a one-time student at Göttingen's Georg-August Universität, liked to compare universities with factories. "It is only advisable," he wrote elsewhere, "to establish a factory if the location, the circumstances, etc. hold the promise of likely profits for the founder; it is just the same for a university."[2]

Similarly, the Göttingen professor Johann David Michaelis explained that the real purpose of a university had very little connection with science or scholarship.[3] Could one possibly believe, he asked, that a "mere love of the sciences" would move great lords and their ministers to fund such expensive institutions? Of course not. The Kammer founded and supported universities for fiscal reasons. By appealing to wealthy students from at home and abroad, successful universities would attract foreign wealth while keeping domestic money at home. At flourishing universities, he argued, one could expect the average student to spend three hundred taler each year. Multiply that by one thousand "polite" students, Michaelis calculated, and a successful university might add three tons of gold annually to the territory's circulating coin.

Michaelis's fiscal university was modeled on Göttingen, the most polite and fashionable university in the Holy Roman Empire.[4] Göttingen's extraordinary success, contemporaries believed, resulted from the work of its own great cameralist, Gerlach Adolf von Münchhausen. For thirty-six years, between 1734 and 1770, he directed the university's daily affairs, governed faculty appointments, kept professors from squabbling, and oversaw the finances. He cared for the university, wrote Göttingen professor Johann Stephan Pütter, with the "tenderness of a father for his daughter."[5]

Münchhausen was not merely curator of the university, he was also the Electorate's leading minister during the 1740s and president of the Hannoverian Kammer after 1753.[6] Moreover, because the elector of Hannover ruled in absentia—the elector had a day job as King George II of England—members of the Kammer and the secret council enjoyed unusual autonomy. Münchhausen was the most important and influential of these officials between 1740, when his brother Philipp became Hannoverian minister in London, and 1770, the year of his death.[7] Like any good cameralist, Münchhausen occupied himself with state finances, promoting and evaluating hundreds of enterprises.

1730s Göttingen, with its fallen manufactures, ramshackle appearance, and miserable trades, posed serious problems for the Kammer. What could be done with the place? Instead of exploiting fields, mines, or forests, Münchhausen turned to the sciences for revenue, hoping that he could use famous professors to draw wealthy "foreigners" into this backward little town. The Kammer had for years worked to attract wealthy foreign criminals to the prison-workhouse in Celle, because they paid more than the locals.[8] The logic in Göttingen would be the same.

When the University of Göttingen first opened its doors in 1734, Curator Münchhausen understood better than anyone that the institution's success depended on the prosperity, appearance and reputation of the town.[9] It was not enough to have famous professors. Other things—suitable apartments, walking paths, coffee houses, pleasure gardens, nice streets, good tailors—were needed to attract wealthy and elegant students. In a backwater town like Göttingen that would not be easy. The Kammer in Hannover, which could do only so much to spruce up the town, needed to ensure that Göttingen and its inhabitants enjoyed some modicum of prosperity. Otherwise, the signs of poverty—abandoned lots, dirty streets, shoddy buildings, beggars—might scare off the elegant students that the university was hoping to attract. Münchhausen even hoped to attract wealthy English students to Göttingen.

"My intention," he wrote to London, "is to draw English lords to Göttingen, where they will fare just as well as in Holland, where they already spend their money abundantly."[10]

The decision to establish a new university in Göttingen created immediate problems. The Kammer hoped that an infusion of money from wealthy students could make the town flourish. But who would convince them to come in the first place? Göttingen's reputation was not the best, so one would have to advertise the place. The authorities, therefore, commissioned a series of works that sang the praises of the idyllic little university town on the Leine river.[11] These quasi-official tributes to Göttingen spoke of its "excellent location, healthy air, good water, and other advantages,"[12] stressing that the "utmost care" had been taken to ensure good police.[13] In 1756 Münchhausen would commission none other than Johann von Justi, whom he appointed as Göttingen's chief police commissioner, to advertise the town.[14]

As Münchhausen strove to burnish the town's image, a few troublemakers wrote less flattering things. One of them was the young Danish student Johann Georg Bärens. Unlike the writers who had been recruited by authorities in Hannover, Bärens did not sugar-coat life in Göttingen. "The weather in Göttingen," he explained, "is not the best." It was too hot in the summer, too cold in the winter, and seemed to rain all the time. The constant rain made for muddy, slimy walking paths. Moreover, there were no decent gardens, and the town was filled with "desolate abandoned lots." Before the arrival of the university in 1734, things had reportedly been even worse. The town had been "indescribably dirty" and "smoky," since half of the town's houses had no chimneys. People flushed smoke through their attic windows after it "had thoroughly seasoned both them and their famous sausages."[15]

Nor did Bärens have any special affection for the locals. "The inhabitants," he wrote, "are basically a coarse, rude, unfriendly lot who cannot, even with the greatest effort, be cured of their bad manners [*Sitten*]." Local burghers were not only lazy and selfish, but they were obstinate too. "They have no understanding of commerce and don't want to learn about it; anyone who has seen Bremen or Frankfurt is considered a well-traveled merchant." Especially remarkable was their "immense hatred of outsiders." New arrivals had often found it difficult to buy food, since the locals would sooner "give it to the pigs than sell it to outsiders." Nor did the town's inhabitants have any notion about "good order or police." The local authorities—judges, mayors, syndics, secretaries, and town councillors—were not much better.[16] Despite his criticisms of Göttingen and his distaste for the local inhabitants, Bärens found two things there that he did like: the university and the local Camelott manufactory.[17] Both institutions had risen to prominence under the watchful eye of Kammerpräsident Münchhausen.

The manufactory owed its success to the efforts of Johann Heinrich Grätzel, who had emigrated to Göttingen from Saxony. Grätzel, wrote Bärens, had relied on "his industriousness, his skill, his understanding, and perhaps also his luck" to build a woolens empire in Göttingen. He had invested an immense sum, more than eighty thousand taler, in buildings, and his manufactory provided work for at least five hundred people.[18] When George II visited Göttingen in 1748, he not only toured the university but also met with Grätzel and visited his manufactory.[19] By 1750 the university and Grätzel's manufactory constituted the twin pillars of Göttingen's prosperity. For Münchhausen, who had patiently protected and promoted both institutions, their continued growth and vigor would be necessary to anchor the finances of the entire region. University and manufactory constituted the twin engines of territorial prosperity. Bärens, for example, even claimed that the university had been founded *in order to* revive the town.[20]

Bärens's report articulated Hannover's concerns about Göttingen much better than any of the fawning, quasi-official tributes to the town. Behind closed doors, the Kammer and the secret council had long worried about what to do with Göttingen. During the 1720s, for example, commerce and manufactures seemed completely stagnant, and the town still bore many scars from the Thirty Years' War. In 1724, concerned about the situation in Göttingen, the Kammer dispatched an agent to observe things and to offer advice about possible improvements.

The 1724 memorandum described Göttingen as a fallen town.[21] Once the regional center of woolens production, Göttingen and its cloth makers had been ruined by a combination of many things: the dissolution of the Hanseatic League, the religious troubles, and the siege and bombardments of the Thirty Years' War (sec. 5).[22] Nor were there many options for reviving Göttingen's productive life. The town was not located along a navigable river or a major road. It had no substantial intercourse with neighboring towns. There were no notable merchants or "capitalists" living there. In short, Göttingen's sustenance and prosperity depended entirely on its own agriculture, brewing, and manufactures (sec. 1).[23]

Though many of Göttingen's residents cultivated small plots of land, Hannover's informant claimed that there were no more than "four or five burghers in the whole town" who could actually support themselves with agriculture, cattle-breeding, and shepherding. He regarded such small-time farming as "imagined sustenance," arguing that it did more harm than good by distracting artisans from their proper work. Göttingen's beer brewing was also a mess. The town's burghers paid too much for bad beer, and even then there was never enough. In short, the town's complicated system of brewing rights needed a complete overhaul (secs. 2–3).[24]

Only through the support and encouragement of manufactures, therefore, could Göttingen hope to recover. Hannover's informant chose to focus on cloth making, which he considered "our best and strongest manufacture." The town's weavers had "to work much and to sell their work." Only in this way, argued the informant, could the town increase its capital and begin to revive after years of decline (sec. 4). According to the memorandum, Göttingen still had about sixty guild masters, eighty journeymen, and one hundred active looms in 1724.[25] The informant estimated that six hundred people depended directly on the manufacture of woolens. Many others like dyers, carpenters, blacksmiths, locksmiths, merchants, tailors, butchers, and bakers benefited indirectly from the trade. In other words, the prosperity of the entire community depended on the success of Göttingen's woolen manufactures (sec. 6). For many years, however, production had suffered from various "jealousies and animosities" among the various guilds of woolen workers.[26] The authorities had, therefore, combined the various woolens' guilds in 1716 to reestablish "harmony" and avoid the constant battles over jurisdiction.[27]

In 1724, at the time of the report, there were several manufactories in Göttingen, and it was not yet clear which of them, if any, would survive and prosper. The oldest one was Lüdeck's royal manufactory, which, as a supplier for the army, had supported many workers over the years. But there were problems. Lüdeck's facility seemed too "limited." It always employed the same number of workers, made the same kind of cloth, and seemed incapable competing with the manufactories in Saxony, which made better woolens for about the same price (secs. 8–9).

Lüdeck had competition from Johann Heinrich Grätzel, a young entrepreneur from Saxony. Grätzel's manufactory seemed more promising. It produced inexpensive cloth and had already attracted new customers, especially among the area's farmers. But, as the memo explained, "this factory is still very small at present." Grätzel's operation had only eleven workers and seven looms, and Hannover's informant guessed that it would never amount to much. Nor could one expect that it would become "a *universal* means for aiding the local cloth manufacture, because it is too difficult for a man like him [Grätzel], who does not have much wealth, to support the work of many."[28] Another of Grätzel's competitors, Rötger Gallenkamp, faced the same problems.[29]

On the whole, then, Göttingen's prospects seemed bleak around 1724.

Its muddy streets were lined with run-down, abandoned buildings.[30] Its geographical isolation hindered major initiatives for trade or commerce. Its inhabitants had no real gift for agriculture or brewing. Worst of all, the town's staple industry, textile production, seemed to be in serious decline. Göttingen's cloth had been banned from the large market fairs in Kassel since 1722, so that many of the local guild masters and their journeymen sat idle. Others were leaving town. Nor did any of the local manufactories seem capable of reviving the town's cloth production. There was, the memo's author suggested, only one solution: the authorities would have to manage Göttingen's cloth manufacture themselves. The Kammer would need to transform Göttingen into a kind of large workhouse, establishing a fund of 4,000–5,000 taler to provide raw wool, tools, and capital to everyone who needed it.[31] The royal government should not, however, allow "one entrepreneur to take the entire work on his shoulders," for such a man might demand Hannover's help in the form of "monopolies, subsidies, prohibitions and other coercive measures, which are a poison to all commerce" (secs. 12–13).

Though Hannover's informant acknowledged that Göttingen's fabrics were "somewhat rougher" than "foreign" cloth (i.e., cloth from neighboring German territories), they were "nevertheless stronger and more durable than the other." The best and only solution, therefore, was to impose duties on cloth from Silesia and Electoral Saxony. Over time, the people would learn to appreciate Göttingen's durable material, and local weavers would improve the quality of their products. Cloth made in Hessen-Kassel, which was inferior to the Göttingen variety, could be banned completely (sec. 14). There was no sense in requiring merchants or farmers to buy local cloth, since the higher prices of foreign woolens would make people purchase more Göttingen *Tuch* (secs. 15–16). Hannover's army, on the other hand, should be forced to use Göttingen cloth for its uniforms. As local weavers perfected their art, this might become "a highly profitable venture for an entrepreneur," since whole regiments had to be outfitted with uniforms and undergarments.[32]

Göttingen *Zeug*—a cheap, carded woolen cloth—was in real trouble. Many of the town's guild members were idle, and the price of *Zeug* had fallen precipitously. Desperate for money and eager to pay their creditors, local weavers had taken to producing thin material, trying to hide missing threads and other signs of their poor workmanship at the fulling mill. As a result, Göttingen *Zeug* had begun to "degenerate and fall into bad credit." Some of the problems were due to external causes, like competition from English cotton and Hamburg woolens. Worst of all, however, had been the 1722 ban on Göttingen *Zeug* by neighboring Hessen-Kassel, which was trying to protect its own weavers. Local producers, explained the informant, felt "as if the world had been nailed shut." Many weavers were ready to leave, and others had "become so lethargic that their work [had] almost come to a standstill." The 1724 memorandum suggested that a lack of proper direction and oversight was also to blame for the deteriorating quality of Göttingen's cloth. Local inspectors, who were supposed to ensure quality, were incompetent and corrupt. Göttingen's weavers seemed to be making and selling whatever they liked, and the inspectors, "sometimes out of ignorance, sometimes out of carelessness, sometimes out of partiality," seemed unable to distinguish good cloth from bad (sec. 20).

It was time, argued the memorandum's author, to institute an "exact supervision of *Zeug* manufacture." Göttingen's weavers had "studied day and night on how to make the fabrics worse, so that they can sell eight or nine or even ten ells per taler." If this continued, there was a danger that all of Göttingen's cloth would fall into disrepute. Since good quality was the bedrock of healthy manufactures, it was critical that the quality of local woolens be improved through "accurate oversight directed by government authority." But Göttingen's weavers also needed the broader support of "good police." It was not enough to set prices for a few staple goods, like meat and bread. Rather, good police should extend to everything that woolen weavers might eat, wear, or use. By ensuring low prices for anything and everything, the authorities could guarantee that quality wares would be available at a "civil price." The hope was that good police would establish a kind of weavers' haven in Göttingen, so that its manufactures would steadily grow and attract skilled workers. Göttingen also needed to improve the appearance of its fabrics. "Since all these things depend on proper direction," wrote the informant, "it would, based on my limited understanding, be very worthwhile if the royal government entrusted a person invested with enough authority, [to oversee] woolen manufactures, [as well as] commerce and police, through which the public could enjoy great advantages at a small cost" (secs. 21–22).

Finally, the 1724 memorandum proposed "external means" for fighting the decline of Göttingen's woolen manufactures. The government would have to be willing to risk some money, it explained. "One has to risk something with all new enterprises." The authorities could impose stiff duties on the so-called "English serges" that were being manufactured in Bremen and Hamburg. Moreover, the secret council had already resolved to establish a market fair in the district of Münden to compete with the one in Kassel. One local merchant was even willing to sell Göttingen cloth in Frankfurt, provided that he got the right support from Hannover. It would also be good, the informant suggested, if "our clothmakers had the courage" to bring their wares to the large "foreign" markets like the one in Frankfurt (sec. 23).

THE RISE OF GRÄTZEL

In the mid-1720s, the Kammer had been trying to revive Göttingen's economy for decades without much success.[33] Hannover wanted Göttingen to provide uniforms for the army, but its various "entrepreneurs," "inspectors," and "factors" had not done so well. Things started to change around 1723 when Grätzel, an opportunistic dyer from Dresden, established his own manufactory in Göttingen.[34] Grätzel was ambitious. He began cultivating woad, and managed to extract a distinctive blue from the leaves of the plant.[35] He also secured royal funds to support his new enterprise.[36] Grätzel intended to restructure the entire system of production and delivery, something that seemed especially threatening to local weavers. Unlike the earlier generation of Göttingen's cloth entrepreneurs, Grätzel was an outsider.[37] He had not been born in Göttingen, and he held no position in the town's government or administration.[38]

Local cloth weavers detested the "foreigner Grätzel," who employed only one of their guild brothers.[39] At the same time, however, the royal government extended its protection to Grätzel. In 1725 Hannover commissioned him to outfit three army regiments with uniforms, for which he received advances of three thousand taler. He also lobbied Hannover for permission to establish his own fulling mill, and he asked for money to support it. The cloth-makers' guild, which considered the establishment of a fulling mill by a "privatus" a violation of its corporate privileges, fought him bitterly. But Grätzel prevailed. By 1726, when local authorities tried to sue Grätzel for unpaid debts, it was already too late, for they no longer had control over him. As a "servant of the crown," Hannover explained, he was no longer subject to local jurisdiction with regard to "debts and other personal matters."[40] By 1730 Grätzel had managed to secure the favor of the crown, largely freeing himself from the impositions of local officials.

Grätzel's manufactory flourished during the 1730s. By 1732 he was employing twenty-eight weavers to make his signature material, Camelott, and he had more than twenty other looms working to supply Hannover's army with red cloth for its uniforms.[41] He also took advantage of Göttingen's public lending house, which provided entrepreneurs with low-interest loans, borrowing eleven thousand taler in 1732 alone.[42] The easing of guild regulations also allowed him to bring in more foreign weavers and dyers after 1731. During this period the authorities in Hannover supported Grätzel at every turn.[43] He convinced them to bend the rules, so that he could borrow eight thousand taler for three years at 4 percent interest.[44] In 1733, the year before the university opened its doors, George II commissioned Grätzel to outfit another five regiments of the Hannoverian army. He was now furnishing the cloth for eight regiments.

Münchhausen was thinking about army uniforms himself as the opening of the new university approached. In an effort to assist Göttingen's cloth makers, he proposed to reorganize the schedule of cloth deliveries to the army. The commanders of the regiments, he suggested, should arrange to have cloth delivered three times a year. The army, in turn, would pay the manufacturers three times each year, instead of only once. With more ready money on hand, Göttingen's cloth manufacturers would no longer need advances from the Kammer. Münchhausen not only wanted to reorganize the regimental delivery system, but he also intended to reward Grätzel, now the biggest single supplier of woolens to Hannover's army.

Grätzel, meanwhile, pressured the authorities for a promotion to "royal commissar."[45] As the inauguration of the university approached, Münchhausen decided to act. "Grätzel has been pushing for the title of commissar for a long time now," he wrote to London. "I have always put him off and urged him to give more evidence of his industriousness." Now, however, Münchhausen felt that the time had come to reward Grätzel. As he explained to London, "this manufactory (Grätzel's in Göttingen) is an important one for the entire land." It would be wise to keep Grätzel happy and to link him more closely to the royal government. The king appointed him "commissar of manufactures" in November of that year.[46]

Grätzel's new position as a royal commissar gave him even greater confidence than before. He knew that Hannover had come to depend on him. After all, no one else in Göttingen had the wherewithal to outfit eight army regiments. Now, as plans for the new university were being prepared, Grätzel became more important than ever to the authorities, for the success of the new university depended on making Göttingen prosperous and attractive. Münchhausen's fiscal university hoped to attract money, but it could not possibly expect to draw hundreds of elegant students to a muddy, ramshackle town.[47] Göttingen had to have impressive buildings, suitable apartments, decent coffee, and clean streets. The success of the new university depended on the prosperity of the town and its residents.

In 1732, as plans were being laid for the new university in Göttingen, Münchhausen was in Celle, some 125 kilometers to the north.[48] The crown and estates were just finishing work on Celle's imposing "prison, work- and madhouse." The prison-workhouse (*Zuchthaus*), which had taken more than two decades to build, was finally completed in 1731 at a cost of one hundred and eighty thousand taler.[49] As Hannover and the provincial estates moved to incarcerate some two hundred "criminals and lunatics" in the new *Zuchthaus* and to establish regulations for its operation, Münchhausen, together with the president of the Kammer, Christian Ulrich von Hardenberg, took the lead. When Hardenberg died in 1732, Münchhausen's responsibilities multiplied. His dream, the dream of the Kammer, was that Celle's prison-workhouse could pay for itself by producing raw materials for other enterprises, like Grätzel's manufactory in Göttingen.[50] Legend has it that the good burghers of Celle, hoping to avoid the disorder and unrest of a university, had chosen instead to be the site of the prison. That is how the Georg-August University ended up in the run-down little town of Göttingen.[51] For Münchhausen, whose authority increased throughout the 1730s and 1740s, prison-workhouse and university were certainly part of the same puzzle. Territorial prosperity, he believed, involved creating the right mix of institutions: universities, manufactories, and prison-workhouses. From the standpoint of the Kammer, these infant enterprises posed similar and interconnected challenges.

Grätzel's manufacturing enterprise was one of Göttingen's major attractions when the university opened its doors in 1734. The *Zeit-und Geschicht-Beschreibung,* an extended advertisement for Göttingen's new university, made special mention of Göttingen's "fine" and "popular" Camelotten, which were manufactured "in all kinds of colors." It also claimed that most of Hannover's troops were outfitted by Göttingen cloth makers.[52] In the years that followed, the Kammer encouraged a virtual building frenzy in Göttingen, and Grätzel became a key ally in the construction effort.[53]

It was also in 1734 that tensions flared between Grätzel and local authorities.[54] Commissar Grätzel, who stubbornly refused to pay local taxes, wanted his freedom from all personal taxation extended for another ten years. He felt that the town fathers had no business demanding taxes from him. He had built many houses. He supported many local weavers. In other words, Grätzel felt that he had contributed more than enough to the town's prosperity, and he complained about having to endure the personal insults and abuse of the local authorities. Hannover sided with Grätzel, expressing its displeasure about his mistreatment.[55]

The royal authorities were so protective of Grätzel because they knew how difficult it was to establish successful manufacturing enterprises. Even as Grätzel succeeded in building up the town's woolen manufactures, Münchhausen was busy trying to cultivate other successful ventures, but his efforts met with constant frustration and failure. At the end of the 1730s, Münchhausen tried to establish cotton manufactures and initiated an effort to revive the territory's linen production. Both projects failed miserably. Münchhausen blamed it on Göttingen's merchants, because he felt they acted "against all reason" in refusing to buy domestic calico and linen. The territory's merchants seemed actively hostile to Hannover's domestic manufactures, and this hostility, which was "without any basis," seemed in danger of ruining many of the costly enterprises that had been established throughout the land. Hannover banned foreign cotton goods from the domestic market in 1739.[56]

At the same time, the Kammer tried everything to encourage linen production in Göttingen, hoping that it would replace foreign cotton goods. But Hannover needed detailed information about the market for cotton and linen. Could enough material be produced to replace the banished cotton goods, or would it be necessary to bring in foreign linen weavers? Were skilled linen printers and dyers already in Göttingen, or would they have to be brought in from outside? Hannover directed the Göttingen authorities to provide answers within two weeks. The Kammer was desperate to find entrepreneurs who could produce linens that people actually wanted to buy. Göttingen's town council responded with an enthusiastic report about the prospects for local linen production, and included linen samples in its report.[57] There were samples from linen producers in Brandenburg, Silesia, and Hamburg. The most striking and expensive fabrics, finely woven and decorated with flowers of red and green, came from Silesia. The cheaper examples included linen cloth decorated with stripes and patterns of green, red, and blue. There were also two samples from local Göttingen linen manufacturers. The Witten sisters in Osterode, for example, had tried to copy printed linen from Leipzig. Unfortunately, their linen was poorly woven, their pinks turned to browns, and their flower patterns smudged. Another local manufacturer, Johann Heinrich Schulmeister, did a little better, but his work still appeared primitive and unrefined compared to the "foreign" samples.

Münchhausen, who examined the linen samples himself, was pleased with Göttingen's report.[58] "Your recently submitted report on printed and colored linen," he wrote, "gave us more satisfaction than all the others that we received on this subject, and the circumstances that you presented give us hope that this manufacture will achieve the desired perfection, just like the woolen mills there." The Kammer was "very inclined" to support "two or three" of Göttingen's fledgling linen weavers. In particular, the dyer Scharff and the manufacturer Schulmeister, whose samples of colored linen were "pretty close to the Silesian," would be eligible for interest-free loans to help them build successful manufactories. The goals of the policy, as Münchhausen explained them, were straightforward: Göttingen's manufacturers ought to produce linen with an "appearance, lustre, and durability" comparable to varieties from Silesia and Brandenburg; and Göttingen's product should be just as inexpensive as the others. The success of the first goal, explained Münchhausen, depended on the "knowledge and good efforts of the entrepreneurs," while the second would take care of itself, since Hannover could impose a duty on all foreign linens.

Despite Münchhausen's best efforts, and despite the thousands of taler advanced to local linen producers by the Kammer, the whole enterprise ended in failure after only a few years. Schulmeister, who always claimed to be making great progress, routinely got cash advances from the Kammer, but the region's merchants never warmed up to his linen varieties. Looking at the linen samples submitted to the Kammer, it is hard to understand Münchhausen's optimism. Schulmeister's dreary, threadbare samples look nothing like the sophisticated, tightly-woven Silesian varieties that he was trying to emulate.[59] It was clearly difficult to know which entrepreneurs deserved support and which did not. One can imagine the appeal of "universal cameralists" like Justi, who claimed to have the knowledge and experience to make such choices. Before 1755, when Justi arrived in Göttingen, Münchhausen used Grätzel as his advisor, turning to him for advice about manufactures, technology, and chemistry. At the end of 1739, for example, when Hannover had been debating the merits of its cotton ban, Münchhausen sought Grätzel's counsel on the matter.[60] Later, in 1750, Münchhausen asked Grätzel to examine a new kind of clay discovered near Lauenburg. The curator wanted to know whether it might be employed like English fuller's earth to cleanse cloth during the fulling process.[61] For the Kammer, Grätzel, who had succeeded where so many others failed, was irreplaceable.

In 1739, two years after the official inauguration of the university, Grätzel informed Hannover about his plans to build a grand new manufactory and dwelling on the Leine river. He wanted a guarantee that the house would be completely free from taxes during his lifetime. The building was to be more than two hundred feet long in the front, and Grätzel had already purchased the plot of land on which it was to be built. Hannover, delighted by the prospect of an impressive new building in the middle of Göttingen, was moved to sing Grätzel's praises to Göttingen's town council.

> You can easily see that the merit of this man, which he has earned with his important manufactures, must make us do everything possible to ensure that his manufactory continues to extend itself and becomes

> sufficiently established through his tireless efforts that his heirs and descendants can continue it. With this in mind, he has decided to build this large house, and we have no doubt that you . . . will do your best to facilitate things there. [62]

The Kammer wanted to know how many building sites there would be and, especially, how Grätzel proposed to improve the appearance of the *Allee,* the street upon which he planned to build. The royal authorities, only too happy to grant permission, even offered help finance the construction.

Grätzel's grand new house on the *Allee* was a massive project.[63] When work began in 1739, the street on which he had decided to build was little more than an abandoned bog. When work was completed in 1745, things looked completely different. The street, now lined with peach and apricot trees, had been drained and renovated. The huge house, with its elaborate portal and impressive pleasure garden, changed the face of the town. Münchhausen could not have hoped for more (figure 3).

Göttingen's local authorities, however, saw things differently. They resented Grätzel's stubborn refusals to contribute anything to the town's treasury. Grätzel had paid no taxes for twenty years. Now, in 1743, he wanted freedom from all local taxes for the rest of his life. Göttingen's representatives were indignant.[64] It was unheard of, they complained, for a manufacturer to enjoy twenty years of tax freedom. In fact, they explained, Grätzel had been in Göttingen much longer than twenty years and had always avoided his obligations. It would be unfair to the community, they argued, to grant him lifelong freedom from taxation. The deputies also explained that "all the guildsmen" expected Grätzel to pay his share. After all, he had "earned a considerable sum" in Göttingen and had acquired many properties and houses, so it seemed only fair that he bear part of the common burden.

Grätzel and the town deputies not only had opposing interests, but they also seemed to hold entirely different worldviews. Local authorities believed that Grätzel, as the wealthiest entrepreneur in Göttingen, owed something to the town and that he, more than anyone, had an obligation to contribute to the welfare of the community. Grätzel, on the other hand, considered himself the wellspring of Göttingen's prosperity. His manufacturing enterprise supported hundreds of workers, and he had invested thousands of taler to improve dilapidated buildings, abandoned lots, and ruined streets. Having given so much to Göttingen already, therefore, Commissar Grätzel felt that he should not be made to contribute even more. He refused to obey the local authorities, acting as though he were subject only to the authority of the royal government.

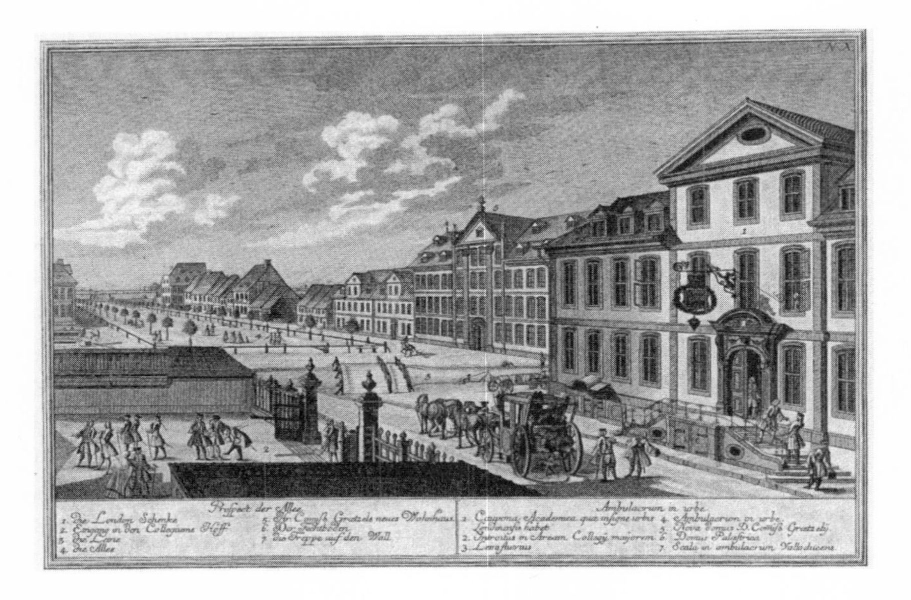

FIGURE 3 View of Göttingen's *Allee* as it appeared in 1747, from a copper engraving by Georg Daniel Heumann. (Heumann, who participated in the public relations campaign for the new university, made the street look more clean and more fashionable than it actually was.) Johann Heinrich Grätzel's house is second from the right, directly next to the *London Schenke*, which is in the foreground. Across the street from the *London Schenke* is the entrance to the university.

Hannover often seemed to endorse that view, stepping in to resolve disputes between Grätzel and the Göttingen authorities. In the 1743 dispute over taxes, for example, Münchhausen essentially ordered Göttingen's town council to grant Grätzel ten more years of tax freedom on his large new house, on his four additional houses, and on any other buildings that he might choose to build or repair.[65] Even the town deputies, he declared, could not deny the benefits that Grätzel's manufactory had conferred upon the town, nor could they dispute the fact that "without him many abandoned building sites would not have been developed."

In 1747, when the dispute between Grätzel and local representatives threatened to get out of control, Münchhausen had to warn Göttingen's authorities to calm down, and he reminded them "how much His Royal Majesty, our most gracious lord, wants to see the further rise of Grätzel's factory promoted."[66] In October, Münchhausen informed Göttingen's town council that the royal government was initiating a special investigation into Grätzel's conflicts with the locals. As he sketched the areas for investigation, however, Münchhausen took one more opportunity to remind the Göttingen authorities that Grätzel had (1) invested a considerable sum to widen and level the town's streets, especially the *Allee,* at no cost to the public; (2) developed many abandoned building sites; and (3) established a manufactory that provided "many residents with their sustenance [*Nahrung*]." The dispute between Grätzel and Göttingen's town council raged on until 1750.[67]

While he was busy alienating most of Göttingen's burghers, Grätzel managed to antagonize members of the university as well. Frau Grätzel contacted two professors, Johann Matthias Gesner and Johann Andreas von Segner, in an effort to find a good tutor for her son. After finding a tutor, Professor Gesner had negotiated the terms of an agreement with Frau Grätzel. The boy would live with Professor Segner (eighty taler annually for room and board), and the tutor, a student named Seidler, would receive a hundred taler annually for his troubles. Johann Heinrich Grätzel initially agreed to the contract. In the autumn, however, he decided to dismiss the tutor without explanation, and he pulled his son out of Professor Segner's house. Then he sent the tutor a few taler and told him to pick out something nice to wear from the manufactory's sample card.[68]

In the weeks that followed Gesner tried to contact Grätzel to discuss the matter, but to no avail. When Gesner formally invited Grätzel to discuss the matter, he received no response; and when they met at a concert in honor of the king's birthday, Gesner complained, "I received a promise from your own mouth that you wanted to visit with me in the near future and take care of the matter." Now, declared Gesner, he could wait no longer. His student, "poor Mr. Seidler," had suffered an injustice, and he felt personally responsible. Gesner threatened to take the matter to the authorities if Grätzel persisted with his "despicable" and "intolerable way of treating us."[69]

In the end, Gesner and Seidler appealed directly to Münchhausen.[70] Gesner confirmed the accuracy of Seidler's story and urged that Grätzel be made to pay the full hundred taler. Grätzel, for his part, insisted that there had been a misunderstanding.[71] The tutor had been incompetent, he explained, so that "my boy did not learn anything for days on end" and "often passed his time on the street."[72] But Grätzel's real defense in this case was, as usual, his own value to town and territory. He regarded Curator Münchhausen's wishes as "supreme commands," and he had contributed "many hundred taler" to town and university. Prorector Ayrer stepped in at this point to resolve the situation, and Grätzel finally agreed to pay Seidler something.[73]

To Göttingen's town councillors it seemed that Grätzel had bought himself the favor and protection of the royal authorities, for he continued to avoid and ignore almost every civic obligation that they tried to impose upon him. Not only did he refuse to pay taxes on his many properties, he even refused to pay for street lighting, the night watch, and waste removal. It was not long before he was embroiled in one more scrape with the town council, this time over the billeting of soldiers. In January of 1745, the Göttingen authorities forced Grätzel to house two officers and fifteen soldiers, whereupon he complained to Hannover that it had caused "all kinds of trouble and disorder" in his manufactory. Münchhausen scolded the Göttingen authorities. After all, they knew that Grätzel's factory buildings and his own dwelling were protected from quartering. They were to remove the billeted soldiers from Grätzel's properties immediately and house them somewhere else. Commissar Grätzel was, Münchhausen stressed, "very busy at the moment making cloth for army uniforms, and he must not be bothered."[74]

Göttingen's community and guild representatives, who could no longer contain their rage and frustration, sent a long letter to Hannover. It read like a list of grievances.[75] "Commissar Grätzel, during his almost twenty-two years here, has failed to contribute anything to the community's funds."[76] Nor had he housed a single soldier during those years, leaving his fellow burghers to bear the burden alone. The town's deputies rejected the claim that Grätzel's work would be disturbed, and they urged Hannover to reconsider, since it was "a hundred times harder" for poor burghers to quarter soldiers than it was for Commissar Grätzel. Moreover, Grätzel had houses all over town, and not all of them were being used for making cloth. He could easily lodge some soldiers in these.

Hannover's orders had forced local authorities to remove the fifteen soldiers and two officers from Grätzel's properties. The town council moved them to a local inn and presented Grätzel with the bill. Grätzel, of course, had no intention of paying, and he once again insisted on his freedom from all local taxes and contributions. The town's deputies begged Hannover, "in the name of the whole burden-bearing community," that Grätzel be forced to pay the innkeeper.[77] Münchhausen responded tersely that now, as before, no soldiers could be quartered in Grätzel's houses against his will.[78] The other matter could be worked out between Grätzel and the mayor.[79]

Göttingen's town council was right to worry about Grätzel's influence, for he had carefully cultivated the patronage of important royal officials. During the early years of the university, when Göttingen was a dingy little provincial town with a lack of adequate housing, Grätzel had provided considerable help.[80] In 1733, as Münchhausen fretted over wood shortages, Grätzel offered to show the locals how to install more efficient stoves for the students.[81] Two years later, he offered Münchhausen his personal assistance, promising to renovate and furnish two dilapidated houses near the Pauliner Kirche for professors and students.[82] And at the official inauguration of the university in 1737, when the Kammer spared no effort to burnish the town, Grätzel again did his part. Deputies from Helmstedt made special mention of Grätzel's efforts. "Factory Inspector Grätzel," they wrote, "had arranged his vestibule like a garden." He had installed a fountain for the occasion. In the background was a miniature mountain decorated with the royal coat of arms.[83]

Grätzel also knew how to buy his way into the learned world. He purchased an expensive cabinet of mineral curiosities in 1736.[84] Bärens reported that the cabinet "surpasses all others of its kind in size and completeness." Grätzel, always looking for a bargain, had snapped it up for only five hundred taler, though it had reportedly cost its owner, a certain Rosinus, sixteen thousand taler to put together.[85] Grätzel installed the collection in a special garden pavilion and opened it for public viewing in the summer of 1737, just as Münchhausen was trying to advertise the new university. The collection gave Grätzel immediate standing in the learned world—the Kaiserlich-Leopoldinisch-Karolinische Akademie der Naturforscher made him a member because of it.[86] More important, it allowed Grätzel to ingratiate himself with fiscal officials in Hannover. He had done much to improve the reputation of town and university. Bärens, for example, considered Grätzel's cabinet of curiosities the only collection in Göttingen, other than the library, worth mentioning.[87]

Grätzel's various efforts to support the new university paid off handsomely. In 1743, his manufactory became a separate juridical entity.[88] His workers—"masters, apprentices, dyers, wool-sorters, cloth-workers, fullers, and generally all those people who work for the factory"—now fell under the jurisdiction of a special "factory court." Workers who spoiled the work because of "negligence or malice" would be punished with "a four-week prison sentence," including two weeks of "water and bread." Repeat offenders would be condemned to one or two years of hard labor. Those who left the manufactory prematurely and without permission could be sentenced to two or more years of "barrow-pushing." Those who encouraged workers to be "unfaithful" could be sentenced to prison or barrow-pushing. The ordinance also provided for special inspectors, "two or three knowledgeable men," who would examine the work of the weavers, identify mistakes, and impose fines for all shoddy work. Inspected cloth pieces would receive an official stamp, and half the money collected from the fines would revert to the manufactory. There were to be no lawyers, "unless absolutely necessary," and all matters would be dealt with in one sitting. Of particular concern were those who came to Grätzel's manufactory, learned its secrets, and then left to work for competitors. Those suspected of such fraud were to be arrested immediately.[89]

Even as Grätzel secured more power than ever over his workers, he began to face competition from two other Göttingen cloth makers, Johann Georg Scharff and Johann Heinrich Funcke. Both were foreign entrepreneurs who had arrived in Göttingen during the 1730s, and both had secured support from the Kammer in Hannover.[90] By 1745, almost all of Göttingen's weavers and spinners were working for Grätzel, Scharff, or Funcke. Scharff's son brought in outsiders from Eichsfeld to teach 120 Göttingen women how to spin wool properly, and he convinced Hannover to let him make Camelott at his fulling mill.[91] When professors began praising the quality of Scharff's cloth, Grätzel must have known he had a problem.[92] When Grätzel complained that Scharff was trying to ruin him, the Kammer looked the other way. According to Bärens, certain "secretaries and other powerful people in Hannover" had decided to support Scharff against Grätzel. Even Gottfried Achenwall, the Göttingen professor of state sciences, was recruited to spread nasty rumors about Grätzel's Camelott. When Grätzel threatened to "leave and to place himself under the protection of the king of Prussia," he was arrested and confined to his house for several weeks.[93]

Grätzel's arrest marked the climax of an intense struggle between Göttingen's three major cloth makers. The Kammer had pursued a complicated policy. Since the 1720s, Hannover had done everything possible to build Grätzel's manufacturing enterprise. By 1745, however, Münchhausen apparently became concerned that they had created a monster. Grätzel had become so wealthy and powerful that he began to pose a threat, so Hannover decided to support competitors like Scharff and Funcke. Grätzel tried to consolidate his monopoly position in Göttingen; when that failed he threatened to leave Göttingen. After investing thousands of taler and decades of effort to build Grätzel's Camelott empire, the Kammer could not risk losing him.

In the years that followed, as Münchhausen built the university, Grätzel worked quietly to consolidate his power. If anything, though, the antagonisms between Grätzel and his competitors grew more intense, and the Kammer finally had to take sides. Münchhausen did not like granting monopolies, whether to manufacturers or professors, but he made an exception in this case. At the end of 1754, as Scharff was threatening to expand his operation, Grätzel secured a royal patent, signed by King George II and countersigned by Münchhausen, granting him a monopoly over all Camelott and *Barracan* production in and around Göttingen.[94] The Kammer had decided that this was the best way keep Grätzel in Göttingen. Henceforth, no one other than Grätzel would be allowed to manufacture Camelott or *Barracan* in Göttingen, or within one mile of the town. Grätzel, in return, was obligated to keep "40, 50, 60 or more looms" in constant activity.

Two months after Grätzel secured his royal patent, three young princes from Hessen-Kassel, grandsons of the king, arrived in Göttingen. Their other grandfather, Landgrave Wilhelm VIII of Hessen-Kassel, had decided to send the princes to Protestant Göttingen in order to protect them from their Catholic father, Prince Friedrich.[95] King George, in turn, decided that his three grandsons should stay in Grätzel's big new house on the *Allee.* It was an enormous honor for Grätzel, who had helped make Göttingen an elegant address, one that would draw well-born students from throughout the empire. With the new patent in his pocket and the three princes in his house, Grätzel was now the unchallenged lord of manufactures in Göttingen.[96]

Though Göttingen was excited about the princes, the princes were not so excited about Göttingen. "For us, it was the center of dullness," recalled the eldest of them, Prince Wilhelm (later, Elector Wilhelm I of Hessen-Kassel). "The formal dinners were awful."[97] Nor were things much better at Grätzel's house. "On August 22," wrote Prince Wilhelm, "there was a small battle at our house. Our landlord, old Grätzel, was the deadly enemy of a certain Scharff. . . . As this one [Scharff] came down the stairs after visiting us, Grätzel had his son attack the man and give him a nasty slap upside the head. Our servants mixed themselves into the fight, and punches flew from all sides."[98] The incident caused a serious stir in Göttingen. Reports were sent to Hannover and Kassel, and the affair was investigated. In the end, young Grätzel was arrested and sent to prison.

The hatreds engendered by these arrests and beatings only intensified the competition between Scharff and Grätzel. The royal patent, however, gave Grätzel extra leverage, and after 1755 he tried to use it as a wedge for driving his competitors and enemies out of business.[99] Scharff and Funcke, on the other hand, tried to avoid the strictures imposed by Grätzel's patent. Scharff, who continued to produce Camelott-like material, began to make it "narrower and shorter" than Grätzel's cloth and to call it "*Concent.*" When Grätzel complained about it, Scharff claimed that he was not producing Camelott. Hannover, however, ordered Scharff to discontinue production. "With regard to so-called *Concent,* which Commissar Scharff presents as different from *Camelott,*" wrote the secret council, "inspection shows that such fabrics have only been given a different name, and that the real purpose is to establish another Camelott factory in town, along with Grätzel's factory."[100] Hannover ordered Scharff to stop making his "so-called *Concent.*" After another appeal by Scharff, Münchhausen had the last word. It had been decided, he wrote, that Commissar Grätzel's privileges would be protected "without any exception."[101]

JOHANN VON JUSTI COMES TO GÖTTINGEN

As Grätzel and Scharff battled over woolens, Münchhausen was considering a mining proposal. Johann von Justi, who had yet to publish any of his major systematic works, sent the *Kammerpräsident* a proposal. Known at this point (1754) mainly for some fluffy literary works, a dispute over monads, and as editor of the *Neue Wahrheiten* (essays dedicated to useful knowledge), Justi had recently left his position as an imperial mining official and professor at the Theresianum in Vienna.[102] Now he was looking for a job. He enclosed one of his essays about a simplified copper smelting process.[103] His letter suggested that the new smelting technique could be of use in the Harz Mountains near Göttingen. If successful, the technique would save considerably on the coal required for the repeated "roasting" of copper deposits. Münchhausen directed two subordinates to evaluate the proposal. The first, a mining official from the Harz town of Clausthal, indicated that Justi's proposal would be impracticable for the upper Harz because of the high sulfur content of copper deposits there.[104] The lower Harz presented a better chance of success. Perhaps, suggested the informant, Münchhausen could send Justi to the lower Harz and offer him a share of the profits from the proposed venture, along with one or two hundred ducats as incentive. But he added a warning: "I communicated with Berghauptmann von Imhoff about this. Based on what he has heard in Vienna he doesn't recommend Justi because the man has fallen into discredit through his excessive boastfulness." The other report from Regensburg reiterated personal concerns about Justi. "He left Vienna mainly due to debts, and because one wouldn't accord him the title "*Bergrath.*" Justi apparently had never gotten permission to leave Vienna.[105]

Münchhausen did not invite Justi to the Harz Mountains, possibly due to these negative reports. But Justi had more to offer than novel smelting techniques. He was writing a book in Leipzig. "His promised system," wrote the Regensburg informant, "will soon show whether his science is merely theoretical, or whether it extends to praxis." This promised system, Justi's *Staatswirthschaft,* would become the most famous and influential of all cameralist textbooks. More important for Justi, however, it would secure him a position in Göttingen.

In April 1755, only a few weeks after the appearance of *Staatswirthschaft,*

Münchhausen offered Justi a position as Göttingen's chief police commissioner (*Ober-Policey-Commissar*).[106] In addition to directing the police commission, Justi received special permission to lecture in the œconomic, cameral, and police sciences at the university. Justi promised to pay special attention to the "fiscal and œconomic lectures." He might give foundational lectures based on his new textbook, offer public lectures on commercial and cameral sciences, or provide special courses on police science. Ultimately, he explained, everything depended on Münchhausen's orders.[107]

Soon after his arrival in Göttingen, Justi began sending reports about the town's police to Hannover.[108] He reported on the poor condition of the night watch, drafted police ordinances to stem student riots, implemented measures to improve the town's wood supply, and made plans to establish a grain storehouse. He also began offering courses. The printed invitation to his lectures posed a direct challenge to the university and its faculty.[109] According to Justi, many academic subjects—metaphysics, philology, and even astronomy—were useless.

> The metrologist [*Meßkünstler*], the astronomer, sees his science, especially the higher aspects of the same, as the most noble and sublime kind of human knowledge. He speaks with rapture about what a stroke it would be for a thinking being if he could at once measure the immeasurable heavens and discover the laws of motion to which the Creator has bound so many worlds. And maybe, if only each successor didn't throw out what his predecessor is supposed to have discovered, one could forgive him his raptures. . . . The metaphysician, when his imagination has penetrated into the primordial corpuscles of substance, into the bond between body and soul—and perhaps even farther—believes that he is engaged in the most splendid science. And he forgets that these are merely his fairy tales . . . and that the origin of corporeal things and the essence of creation will always remain hidden from human knowledge.[110]

A consistent commitment to useless sciences like metaphysics and higher astronomy, Justi argued, had generated sweeping disdain for universities and professors, so that the learned (*Gelehrten*) generally failed to obtain important posts in the Kammer. Rather, it was the *Ungelehrten,* those onetime "footmen, flatterers, clerks, ordinary hunters, petty tax collectors and the like," who rose to the most important and lucrative state offices. But one should not blame the sovereign or his ministers for the success of the *Ungelehrten.* "My God! What could a regent or a prime minister do with the most profound metaphysician, the greatest metrologist, the most famous astronomer, or the most rigorous philologist in the business of state?"[111]

Justi's attacks on academic uselessness and pedantry carried significant moral overtones. Profound metaphysicians and astronomers, caught up in fictitious worlds of their own making, were guilty of neglecting their duties as state officials. For Justi, useless science was at bottom selfish science, living parasitically, like the common idler or drunk, from the labor of a productive society. Göttingen's useless professors, alone in their studies and divorced from the productive society that sustained them, created fancy systems of no interest to anyone but themselves. They deserved expulsion from the university. Metaphysicians, philologists, and astronomers were the beggars, gamblers, and drunkards of the academic world.

Justi blamed academic uselessness on the legacy of German universities. "Scholars once seemed to constitute a kind of special kingdom separated from civil society. They worked for themselves. They shared their thoughts and discoveries only amongst themselves, and since they worried more about fame and the esteem of their fellows than about the benefit of the people . . . they wrote in a language wholly unfamiliar in civic life."[112] He thus proposed an entirely new "faculty of œconomic sciences."[113] Like the other professional faculties—law, theology, and medicine—the new faculty would demand large numbers of professors to teach "auxiliary sciences" such as mineralogy, chemistry, forestry, and accounting. The new faculty would train fiscal officials and offer "advice for the many institutions and undertakings of the state, for which one must often turn to foreigners at great expense."[114] For Münchhausen, who had wasted thousands of taler on bad entrepreneurs, this had to be appealing. Where Grätzel had attacked the privileges of Göttingen's guilds, Justi now proposed to challenge the monopoly over professional education held by the higher faculties. His manufactory of knowledge would enlist the sciences in the service of the Kammer. It was the same vision that Heynitz realized at the Bergakademie Freiberg a decade later.

Justi had also been brought to Göttingen for other reasons.[115] Münchhausen hoped to enlist him in the effort to improve Göttingen's police and reputation. Despite the Kammer's best efforts to generate good publicity, reports of high prices, dirty streets, empty lots, unreliable mail, and foul water persisted.[116] Justi not only aimed to reform the university, he planned to clean up the town.[117] He would bridge the gap between town and university, artisan and professor, science and practice. His administrative efforts in Göttingen—what he actually did as police commissioner—have not received much attention.[118]

Justi soon brought good police to Göttingen's streets. "I have arrested Johann Julius Kretzer," he wrote Hannover in December 1755, "because he dared to have an alms certificate [*Bettelbrief*] prepared for him." The letter was a forgery, and Kretzer had used it to collect contributions from unsuspecting students: "He has collected more than seven taler in this way." Kretzer's mother, notorious for her "pimping" [*Küppeley*] and living in Göttingen without permission, was also implicated. Though the police commission had initially decided to send Kretzer and his mother out of town, Justi thought it might be better to make an example of them. When he turned to Hannover for orders about how to proceed, he received a sharp rebuke. "In the future, if someone is to be arrested this should not be arranged by you but by the authorities." Justi was to leave arrests to the bailiffs for the poor [*Armenvoigte*], whose job it was to arrest street beggars.[119]

Justi was indignant. He wrote a long and despairing letter directly to Münchhausen. "I must humbly confess that the content of this order has distressed my thoughts very much." According to Justi, section 23 of his "Instruction" granted explicit authority to arrest all "suspicious riffraff, vagabonds, unknown beggars, gamblers and other people engaged in similar unlawful occupations."[120] If he were denied the authority to arrest, his position would be undermined.

> The power of arresting is so essentially and inseparably connected with a police directorate . . . that it could not be taken away from me without making the exercise of my office completely impotent. Even an ordinary police commissioner or inspector must have this authority if he is not to be an inferior servant of the town council. Otherwise any police assistant or functionary could carry out the same things that a police directorate or commissariat is established to do. The matter speaks for itself, and the police principles in all books that address it are so far unanimous on this point that Göttingen would have the first example of a police directorate or commissariat deprived of this power.[121]

Justi sensed a plot. "The alacrity with which I have attacked police affairs," he explained, "has earned me enemies, and there will be no lack of cabals, intrigues and slanders against me." Everything depended upon Münchhausen's favor. "Your Excellency," Justi wrote, "is master and lord not only over my authority to arrest, but also over all the other offices which have been entrusted to me." He begged Münchhausen to protect him, claiming that any further loss of authority would "banish" him to the lecture room.[122]

Like Grätzel before him, Justi now became embroiled in a nasty struggle with Göttingen's town council. He almost came to blows with the secretary of the town council, Justus Claproth. Claproth had neglected to put the word *Herr* in front of Justi's name in the minutes of the police commission meetings. Justi complained about it, instructing Claproth to add *Herr* to the protocols, but Claproth refused.[123] Justi then complained to Hannover that "those who take the minutes at the town council here, in order to belittle me, never use the word 'Herr' when they refer to me in the protocols; instead, they merely write "Bergrath von Justi," despite the fact that they always, without exception, place the title 'Herr' in front of their mayor and syndics." In the case of a local body, like Göttingen's town council, Justi argued that this constituted real injury to his name and honor.

Justi complained about the repeated snubs, but nothing changed. In fact, it got worse. Claproth now had the gall to put *Herr* before Assessor Insinger's name. "I therefore informed Secretary Claproth, through a police assistant, that he should be so good as to add 'Herr' to my name in the future," explained Justi, "because I would otherwise consider it an injury." Claproth, he explained, should have thanked him for the "friendly reminder." Instead, Claproth sent an "unbecoming letter," which Justi promptly forwarded to Hannover. "It alienates me exceedingly," he wrote Justi, "that you deemed to give me orders verbally through a police assistant."

> I feign to know as well as you do what belongs in the protocols, and what does not, so I will follow my own inclinations. Perhaps you want me to follow your orders. I must tell you very directly, that I will certainly not do so, because I in no way consider you my superior, and the royal government never indicated that the council members would be commanded by you. By way of information, be aware that I will send a complaint in the next mail about your unseemly conduct in having me reprimanded through a police assistant. [124]

As promised, Claproth wrote a letter to Hannover about the affair.[125] Justi's assistant, Hildebrandt, had come to his apartment "yesterday afternoon at about 3 o'clock," while he was entertaining guests. Claproth worried "that I have become a laughingstock. And I have additional disadvantages to worry about, since I have a considerable connection to the students because of various private lectures. How easily could this story, already known throughout the whole town, contribute to a disadvantageous contempt for me."

Despite Claproth's complaint, Justi eventually got his way. The secretary was directed to place *Herr* in front of his name in the protocols. But Justi had clearly tried Münchhausen's patience. "His Excellency remarked that he would have preferred to see the Herr Bergrath deal with Secretary Claproth's insinuations in a different way, thereby saving himself the vexation and waste of time."[126]

Less than a month later, Justi and the town council came into conflict once again. "Ten days ago," Justi wrote to Münchhausen, "the second police assistant Hildebrandt died from drinking too much schnapps."[127] The previous police assistants had been so bad, so completely incompetent, Justi explained, that he could not use them at all. When the last one died, Justi had hoped to fill the position with someone useful, but the town council had appointed "the worst of all the candidates." The new police assistant proved even worse than the old ones, "since he sits day and night at the gambling tables, as is well-known." Now, Hildebrandt's death had left another vacant position, and Justi hoped to fill it with a useful assistant. Once again, however, the town council intended to vote on it immediately without consulting him at all. "I humbly request," Justi wrote Münchhausen, "that your Excellency order the town council, per rescript, to send me the names of those who have applied for the position, and to reject those whom I deem unsuitable."

Münchhausen did not order Göttingen's authorities to remit the names directly to Justi. Rather, he ordered the town council to send its list of proposed candidates to the Kammer, which then sent it back to Justi in Göttingen, who in turn sent it back to Hannover with the following remarks:

> #3 Johann Heinrich Kleinschmidt: is an idle and lazy person.
> #8 Johann Georg Weitekamp: is too simple and awkward.
> #10: Johann Wilhelm Lütje: is consumptive and sickly.
> #11: Tobias Riemenschneider: is given to drink, lazy, simple, and completely unqualified.
> #13 Georg Andreas Kaufmann: is reputedly an idle and lazy man.
> #15 Daniel Polle: is very attached to schnapps and is always lying around the pharmacy. Moreover, he probably has the most supporters, since the town council already gave him the interim construction assistant position, which the deceased Hildebrandt also had.[128]

Three days after Justi mailed his remarks to Münchhausen, the Kammer informed the town council that several of its candidates were unacceptable. "You will," wrote the ministry, "not consider Joh. Heinr. Kleinschmid, Joh.

Georg Weitekamp, Joh. Wil. Lütje, Tobias Riemenschneider, Georg Andreas Kaufmann and Daniel Polle at the upcoming vote."

In 1756, Justi extended his policing to the university. He complained that a band of unruly students was terrorizing the once quiet neighborhoods around Göttingen. Led by two students, la Pic and la Peine, the band had insulted the municipal guard and abused "honest people" in the streets at night. The group had passed through Justi's own neighborhood, stopping in front of Amtmann Hiepen's house "this past Sunday night between 10 and 11 o'clock," and shouting shameful things about *mademoiselle* Hiepen. They stopped in front of Hiepen's house again the next night singing "the most shameful songs." And they came through again the night after that "screaming, shouting and singing in the streets." Though he had no direct jurisdiction over university students, Justi felt that his orders to maintain "the nightly peace and order" warranted his direct involvement in the matter. Accordingly, he urged Prorector Ayrer to expel la Pic and la Peine from the university and to remove them from town.[129]

While he was arresting beggars and chasing down rowdy students, Justi also founded a periodical, the *Göttingische Policey-Amts Nachrichten.* He dedicated it to the improvement of the *Nahrungsstand,* a term he used to denote society's productive classes—its miners, farmers, manufacturers, merchants, and shopkeepers. It was the central organizing category of his police-cameralist program.[130] Justi's well-ordered police state was divided between *Nahrungsstand* and *Gelehrtenstand,* between those who labored and those who directed. As an organ of good police, The *Policey-Amts Nachrichten* aimed to serve as a mediating instrument between these two realms. The periodical was an embodiment of Justi's cameralist vision, juxtaposing chemical treatises and police ordinances, botanical essays and price tables. Essays about lixivial salts appeared next to police ordinances about vagabonds. Treatises on smelting and assaying accompanied the weekly prices for bread and carp.[131]

Despite his best efforts, however, Justi did not secure state funds to support the *Policey-Amts Nachrichten.*[132] Nor did he have much success with other projects. Göttingen's Royal Society of Sciences, founded by George II in 1751 and then under the leadership of Swiss anatomist Albrecht von Haller, rejected his proposals.[133] The town council also stubbornly resisted his projects. Justi may have written the seminal work on police science, but he was a very bad police commissioner. When he left Göttingen in 1757, after the outbreak of the Seven Years' War, it was unclear whether he was fleeing the Austrian troops or personal debts.[134]

Justi went to Copenhagen, where he found a powerful patron in the Danish minister Johann Hartwig Ernst von Bernstorff. When they met, Justi offered to travel around Denmark and write a book about manufactures, at the government's expense.[135] He dedicated the book to Bernstorff, who had shown him "so many kindnesses" during his stay in Copenhagen.[136] Not everyone in Denmark was as kind. A Copenhagen correspondent named Daß, for example, complained that "we have gotten a new German, the mining councilor Justi, who, after so many changes in his studies, his profession, and his religion, had finally become the head of police in Göttingen."[137] According to Daß, Justi had left Göttingen to escape his creditors and his Catholic wife.[138] Now that he was in Denmark, the letter continued, Justi had already turned up at the Royal Society of Sciences in Copenhagen, where he had "read a lot of claptrap about the three principles—water, oil and quicksilver—which are supposed to reveal and explain all the secrets of nature and even lead us to the discovery of new mines. Of course, what won't a German promise (actually every windbag and fortune-seeker, even the English, as the latest events teach us) for wonderful money!" Daß also reported that Justi, who had two wives, had written a piece on the "usefulness of polygamy."[139]

GÖTTINGEN AFTER JUSTI

When Münchhausen looked to replace Justi after the end of the Seven Years' War, he looked to Göttingen's most influential professors for advice. Johann David Michaelis urged the curator to hire a Swedish-style Linnaean œconomist. Too many of the German œconomists, he argued, were useless projectors like Justi.[140] Münchhausen initially went a different direction, granting Johann Cristoph Erich Springer, an official from Ansbach (in Franconia), permission to offer lectures in the "œconomic, financial, and police sciences." When Springer began lecturing in May 1766, Münchhausen ordered Professor Johann Stephan Pütter, a jurist, to submit detailed reports about his courses.[141] In Pütter's opinion Springer had made two major mistakes. First, he had transcribed his lectures and read them word for word; second, his lectures on financial and cameral science were redundant with other lectures, especially Achenwall's lectures on politics. "Mr. Springer," Pütter explained "would presumably have found more applause had he from the beginning focused more on the actual œconomy of the land: cultivation of the earth, fields, meadows, forests, mines, etc. along with cattle breeding."

Münchhausen then turned to Michaelis for a second opinion. "I answer reluctantly the questions submitted to me by your Excellency, especially the question regarding Mr. Springer," he replied. "Since your Excellency orders it, I must put aside the fear of insulting someone, but request that this first part of my answer will not become part of the official record, and will rather be discarded or graciously returned to me."[142] Michaelis considered Springer incompetent. He was ignorant of forestry, weights and measures, botany, and means of taxation. "We spoke of forestry in the Harz," wrote Michaelis. "In his opinion, it was very bad." Springer was completely ignorant of simple things. He did not know what a larch tree was and seemed equally stupid about royal patents, the army, domain lands, and tolls. Springer, in short, was ignorant of the things "which surely every beginner in *Landes-Oeconomie* knows." "About œconomy," Michaelis complained, "he understood theoretically even less than I do, and I of course do not hope to become a professor of œconomy." Springer was worse than incompetent. He was, argued Michaelis, the wrong *kind* of cameralist, focusing on fiscal-juridical issues rather than on œconomy proper.

> His so-called "Camerale," in which he still wants to derive everything from the fisc, something like the Wolffians once did, is (so far as I can tell from his prospectus), not œconomic, but in its way juridical. Since the general part already appears in other law courses . . . this seems to me an unnecessary course that only robs the enrolled students of their time. Such reproduction of courses appears altogether harmful. I also think that the law professors lecture so well that we don't need his *Juridico-Camerale*. . . . This only I know, that they [Springer's lectures] are not œconomic, and if Herr Springer wanted to lecture for 100 taler, the acquisition would be too expensive for me. Such a professor, who is not at all known through his writings, even if he were better, does not contribute to the fame of the university. The worst is that, as soon as he was a professor, the hope to get a real œconomist for the university would become that much smaller. For the 700 taler or 800 taler that he is asking, one could maybe get Schreber, who is a very different man.[143]

Other critics echoed Michaelis's opinion that Springer's cameralism was of the fiscal-juridical variety. "Though Herr Michaelis's account is a bit excessive," wrote one anonymous reviewer, "there is presumably something true in it."[144] Springer, the informant continued, also lacked requisite understanding of important "auxiliary sciences," like chemistry and botany.

After Springer's failure, Michaelis renewed his call for Swedish-style œconomy in Göttingen. It was crucial, he argued, that the new candidate understand Sweden's "scientific œconomy," which was "so closely and naturally tied to botany."[145] He proposed the Swede Johann Andreas Murray, a young Göttingen professor of botany, who had studied with Linnaeus. At almost the same time, Professor August Schlözer proposed another young candidate with Swedish credentials: Johann Beckmann, a former Göttingen student who was studying with Linnaeus in Uppsala.[146]

The ministry opted for Beckmann, who arrived in Göttingen during the autumn of 1766. In the spring of 1767, less than a year after his arrival, Beckmann advertised his intention to offer lectures in botany during the summer semester. Upon learning of Beckmann's intentions, Murray, the professor of botany, wrote Münchhausen a despondent letter:

> Your Excellency was gracious enough to recommend botany to me, a subject which I have now practiced for so many years—with, as I hope, approval—and with the understanding that I might see my fortune furthered in this way. It can obviously not be indifferent to me that another should reap the fruits of a science that has required the application of so much effort and so many costs with so little income, a science which I cannot employ outside of the university. I have been assured that this step of Mr. Beckmann's occurred without your permission, and actually against your grace's will.[147]

Beckmann's published intention to lecture in botany posed a considerable threat to Murray, whose survival depended upon the success of his lectures. Murray, who feared that Beckmann would draw students away from his courses, charged the new œconomist with incompetence and dilettantism. "Botany," he railed, "is too important a science for someone to learn it from a flimsy schoolbook, with a barely six-month stay in a place where one does not understand the language. . . . And I know Mr. Linnaeus too well, to believe that he would reveal the most difficult aspects of his science—insofar as this were even possible—out of mere politeness to a stranger."[148]

Münchhausen, a confirmed opponent of monopoly, valued competition among professors, but he worried that Beckmann might be infringing on the privileges of the medical faculty. So he asked the head of the medical faculty, Richter, for an opinion. "I can't see how Mr. Beckmann proposes to read in botany, a part of medicine, without infringing upon our rights," Richter complained. "I am amazed that he has already begun to strike out around himself, especially since no one has heard him or can judge his skill." He insisted on a strict reading of the university's legal privileges. "As worthy as it seems to support good enterprise, and to avoid monopoly, this rule cannot interfere with the lawful enforcement of our rightful privileges."[149] The medical faculty controlled the botanical garden, and Richter made it clear that Beckmann would have no access to the plant species he needed for his lectures.

Münchausen informed Beckmann that his proposed lectures in natural history would not be allowed due to the opposition of the medical faculty. Beckmann was incredulous, and he threatened to offer the lectures with or without the approval of the medical faculty.

> The medical faculty cannot rightfully protest, nor will it protest in actuality. My intention is not to teach botany medically . . . but only insofar as it is a part of natural history, physics and œconomy and has been read, not only at all universities by physicists, œconomists and cameralists but also in Göttingen. . . . I also know from members of the faculty that my lectures will not alienate them . . . since, as they would freely admit, mineralogy and chemistry, which can also have a connection with medicine, are already read by members of the philosophy faculty. Moreover, my listeners in natural history have been mainly gentlemen of nobility, jurists and theologians [and not medical students].[150]

It did not matter. Three days later, Beckmann learned that the medical faculty had rejected his botanical lectures. "Though you promise not to read your botany medically," Hannover informed him, "the subject does not really admit of boundaries."[151]

But Münchhausen was not about to let the medical faculty get in his way. In 1767 he authorized Beckmann to offer lectures in œconomy during the winter semester. The proposed lectures were largely the same, but botany was nowhere mentioned. Moreover, seeing that there was no way for Beckmann to gain access to the botanical garden, Münchhausen established a new "œconomic garden" for his lectures.[152] By 1768 Beckmann was offering his botanical lectures, only now they appeared under a different name: *Oeconomia.*

In May 1770, about six months before Münchhausen's death, George III's royal patent declared Beckmann "our ordinary professor of œconomy."[153] Beckmann's œconomy enlisted Linnaean description in the service of good police. His sweeping classifications of agriculture and technology, which would make him famous, aimed to categorize the *Nahrungsstand,* that great disordered sphere of productive activity. As Beckmann explained it, "the knowledge of trades, factories and manufactories is indispensable for anyone who wishes to apply himself to the police and cameral sciences. For one should at least be familiar with what one intends to arrange, design, direct, judge, govern, maintain, improve and use."[154]

Twelve years after Münchhausen's death, in 1782, Johann Beckmann had established himself as Göttingen's ordinary professor of œconomy. He had made a name for himself throughout the German-speaking world with his systematic works on the agricultural and technological sciences.[155] Students came from all over the empire to hear his lectures. When Hessen-Kassel offered him a prestigious position as fiscal councilor (*Kammerrat*), Hannover found itself forced to respond with a salary increase.[156] By 1784, Beckmann had been appointed *Hofrat* by special order from London.[157] He had become a valuable piece of academic property.

CONCLUSION

Johann von Justi was a smooth operator who fooled many powerful people through the years. Münchhausen was not one of them. Hannover's sly old *Kammerpräsident* knew exactly what he wanted from Justi. Unlike Maria Theresa before him and Frederick the Great after him, Münchhausen was not taken in by Justi's mineralogical and metallurgical fantasies. Justi, you will recall, had proposed a simplified copper-smelting process for the ores of the Harz mountains. After checking with officials in the Harz, Münchhausen abandoned that proposal, but he did not abandon Justi. Instead, Göttingen's curator invited the *Bergrath* to Göttingen, provided him with a position as chief police commissioner, gave him permission to lecture at the university, and directed him to advertise the town.[158]

Münchhausen valued Justi not for any "practical knowledge" about mines or town policing, but for reasons of reputation and publicity. The curator, who knew from his informant in Regensburg that Justi had an important book in the works, brought him to Göttingen just before it appeared. Münchhausen had a habit of collecting talented professors to build the reputation of his university. The book in question was Justi's *Staatswirthschaft,* the most successful cameralist textbook ever written. It would make him famous. By 1755, when *Staatswirthschaft* appeared, Justi was already in Göttingen giving lectures on the cameral sciences, publishing his *Policey-Amts Nachrichten,* and running the town's police commission. Though Münchhausen supported Justi's various efforts at publication and publicity, Hannover rejected all the police commissioner's efforts at real change. The Kammer did not want him policing the streets, arresting people, or initiating ambitious new projects like the grain storehouse. Münchhausen expected Justi to burnish the town's image, but not the town itself.

By the time Justi arrived in Göttingen, Münchhausen had already devoted decades to improving Göttingen's appearance and economy. Central to that effort had been the strategy of reviving the town's woolen manufactures. After several missteps, Hannover finally discovered an entrepreneur from Dresden named Grätzel. With the help of the Kammer, Grätzel had built a small woolens empire in Göttingen, and he had used the profits from that enterprise to improve the town itself, building houses, draining swamps, displaying cabinets of curiosities, and making things pretty for the university's elegant students. Despite bitter opposition from the locals, Hannover resolutely supported Commissar Grätzel, even when he and his family did stupid things like firing tutors and starting fistfights. Münchhausen understood Grätzel's importance in a way that local authorities and university professors did not. Unlike academic cameralists, he knew how rare it was to find a successful entrepreneur who could produce attractive cloth with local weavers. Unlike mayors and syndics, he valued well-woven fabrics and impressive new buildings more than tax contributions. His was the standpoint of the Kammer, a perspective that considered knowledge useful insofar as it could be sold. Justi's cameral sciences had value to Münchhausen not because they would make the mines more efficient or the town better ordered, but because they could build the reputation of Göttingen and its university.

CHAPTER FOUR

The Cameralist and the Ironworks

Johann von Justi, the great bard of the Kammer, died poor, blind, and alone in a fortress overlooking the Oder River.[1] Frederick the Great imprisoned him there, in Küstrin, for fraud and mismanagement. The king had been confined there himself four decades earlier by his father, Frederick William I. The old king thought that Frederick needed more discipline and better knowledge about everyday fiscal administration. In 1768, after Justi wasted more than one hundred thousand taler on an elaborate enterprise in the Neumark, Frederick probably thought the same thing about him.

Three years earlier, in 1765, Justi had received a commission as Prussian Berghauptmann and royal leaseholder (*Pächter*), securing an agreement to administer a sprawling ironworks in the Neumark, east of the Oder.[2] He made an immediate impression on Frederick: this was not another plodding, dependable domain administrator; this was not a man who spoke only about pigs and bees, trees and fields, bookkeeping and economizing; this was not his father's cameralist. Frederick knew that world well, having served hard time in Küstrin, where he was force-fed on a strict and very dry diet of tenancy litigation, account audits, soil cultivation, and forest management. "The crown prince's life in detention was simple and monotonous. For an eighteen-year-old it was deadly dull, but it was as useful as it was difficult. In the daily routine of the Kammer, the future ruler saw for the first time the workings of the economy and the fiscal system, the nature and content of contracts, and the operations of the administrative machine.[3]" The proud fortress town of Frederick's youth was barely recognizable in 1768, when Justi was jailed in the *Stockhaus* there. Once a bustling center of provincial government and trade along the Oder, the city, like the fortress, was now a mere shell of its former self, a fitting end for the old cameralist, now blind, disgraced, and impoverished.

Many have speculated on Justi's guilt or innocence, but they have had little success in reconstructing the events that led to his imprisonment. Some have scoured the published record looking for answers, without much luck.[4] Others have made excuses for Justi: he was a victim of political intrigue, of vindictive bureaucrats, of his blindness, or of his own scheming secretary, Virmitz.[5] These excuses mirror the endless defenses mounted by Justi and his family after he was arrested. Only Ferdinand Frensdorff really dug into the archival record, and that was more than one hundred years ago; he has had the last word on the subject ever since. Frensdorff was so thorough and so authoritative on the subject—even the *Acta Borussica* referred to his account as exhaustive—that few have bothered to venture back.[6]

Fascinating and thorough though it was, Frensdorff's biography revealed little about Justi's failed administration in the Neumark, relying entirely on records from the Prussian Secret Archive in Berlin,[7] together with published material, to reconstruct the story. These documents offer fascinating glimpses into the disputes that erupted between Justi and the Kammer in Küstrin, the fiscal-administrative heart of the Neumark.[8] But Frensdorff did not use another, more extensive set of documents in the Brandenburg State Archive, which detail the planning, administration, and eventual collapse of Justi's ambitious metallurgical venture; nor did he pay much attention to Justi's extensive chemical and mineralogical writings.[9] These documents provide everything that Frensdorff could not: a better sense of Justi's ambitious iron and steel projects; detailed accounts of his struggles with local foresters, royal inspectors, and provincial finance councillors; and evidence about how Justi used his published writings to further his administrative ambitions.

Provincial officials in the Neumark considered Justi's administration a disaster in almost every way. He was reckless; he mismanaged everything; he could be overbearing, dictatorial, and sometimes violent; he seems deliberately to have defrauded the royal treasury; and he was virtually blind toward the end. In short, when it came to actual administration, Justi violated almost all of his own well-known cameralist principles.

PUBLIC SCIENCE AND PRIVATE INTEREST

In the summer of 1765, Justi finally secured the position that he had long coveted, receiving a commission as a Prussian Berghauptmann, with direction over all royal iron and steel production in the kingdom. After years of itinerant work—lectureships in Göttingen and consulting in Denmark—he was now a well-paid Prussian official.[10] The Kammer in Küstrin got official instructions from Berlin in November of 1765. The king had resolved to establish a sprawling new iron and steel complex east of the Oder. Berghauptmann von Justi, already famous for his many books, would now assume control of the existing ironworks in Vietz and Kutzdorf; what is more, he promised to build a new steel and ironplate works along the Zanze, a small river some forty kilometers east of Vietz. Justi assured the king that his proposed "combined works" in the Neumark would free Prussia from dependence on foreign (that is, Swedish, Saxon, and Siberian) iron and steel.[11]

The existing ironworks complex included a blast furnace in Vietz, two hammer forges in Kutzdorf, several smithies, ironstone mines, and large swaths of forest for charcoaling. Justi guaranteed that he could make the existing works more profitable than "Entrepreneur Winckelmann," the previous tenant. Instead of offering a greater annual lease payment than Winckelmann, however, Justi promised to reinvest the additional surplus into the new royal works on the Zanze. In addition to his annual salary of two thousand taler, Justi demanded that the new works on the Zanze remain rent-free until construction had been fully completed. Simply put, Frederick agreed to bankroll the new enterprise, and Justi pledged to administer it. In practice, however, it was more complicated than that, because Justi assumed two roles. He was on the one hand a Berghauptmann, a royal administrator charged with managing all Prussian iron and steel production; he was, on the other hand, a tenant-entrepreneur (*Pächter*) with contractual rent obligations to the crown. Justi's own writings suggested the potential conflict involved in such a position.[12] As a royal administrator, he was supposed to have only the royal interest and the general welfare as his guides; as a tenant-entrepreneur, however, Justi would naturally look to his private interest. In other words, because he was both a royal Berghauptmann and a leaseholder, Justi would have to supervise himself.

Frederick's commitment represented a significant gamble: by investing large amounts of money, he intended to create a domestic industry that could supply all of Prussia's needs for steel, blackplate, and tinplate.[13] The immediate point was not to create an export industry—though Justi promised that too—but to supplant the tons of expensive foreign steel, iron, and tinplate that Prussia imported from Sweden, Saxony, and Russia every year. In Frederick's opinion, the trade in iron and steel was draining money from the kingdom at an alarming rate and at a time when he needed revenue more than ever. In private audiences with the king, the old cameralist convinced Frederick that he had the solution. And why not? Justi's writings offered a blueprint for success.

We will never know just what Justi said to Frederick in those private audiences, but his chemical writings suggest how he used the sciences to sell his iron and steel projects in the Neumark. By the time he pitched his enterprise to Frederick in 1765, Justi had been working on chemical questions for about ten years. But he achieved his greatest notoriety in chemical matters during the early 1760s, when his dispute with Johann Heinrich Pott, a member of the Berlin Academy of Sciences, turned nasty. The substance of the dispute centered around Pott's distinction between gypsum and calcareous earth, a distinction that Justi found objectionable. But Pott's classification system did not explain the vitriolic nature of Justi's criticisms. He aimed to gain favor with academicians in Berlin, and he hoped to establish himself as a chemical expert. Pott, who was just then falling out of favor in the Berlin Academy, made the perfect target. It was the same strategy that Justi had employed some thirteen years earlier with his controversial essay on monads.[14]

Justi used the first two volumes of his "collected chemical writings" (1760–61) to consummate the attack on poor old Pott.[15] He ridiculed the notion that someone "who has only assayed the gypsum of a single quarry," could presume to reclassify all the rocks and types of earth in the entire world. It was an outrageous accusation, especially since the conclusions in Pott's *Lithogeognosia* reputedly rested on more than thirty thousand experiments. More damaging for Pott, however, were Justi's claims that he was the wrong kind of scholar, given to small-minded revenge and incapable of accepting criticism. This attitude, complained Justi, "is proof enough that he can abide no contradiction; because of my love for humanity and my keen desire for the increase of the sciences, it is a disposition that I hate to see." Useful scholars, he continued, had to accept their own fallibility. Those who did not ignored the "duties of society." One could not conduct a useful dispute with men like Pott.[16]

Justi's chemical writings were strategic in other respects as well. The first volume alone, compiled and prepared in Berlin during 1760, included essays on ironstone, tinplating, and steel production; the second volume featured a treatise on "the smelting and preparation of good iron."[17] In other words, Justi was already promoting his ability to transform Prussia's "iron economy" five years before receiving his commission as a Berghauptmann. Moreover, his work fused cameralism with chemistry, creating a hybrid discipline. He was, on the one hand, careful to reference major scientific authorities—Georg Ernst Stahl, Johann Joachim Becher, Étienne François Geoffroy, René Antoine Ferchault de Réaumur—and he constantly emphasized the importance of good chemical principles; on the other hand, his chemistry was so suffused with fiscal-police principles that it would be arbitrary to distinguish it from his larger cameralist project. In Justi's hands, that is, chemistry *was* a cameral science.[18]

Consider Justi's essay on iron smelting, in which he claimed that iron is not present in iron ore, or ironstone, but that it first arises during roasting and smelting.[19] Though Justi admitted that the claim seemed paradoxical, he assured his readers that it rested on "completely unshakeable principles, which constitute an almost demonstrable certainty." Moreover, the essay, which took the form of a letter, was based on an audience with someone identified only as "your excellency"; a footnote explained that the essay "had initially been presented in writing to a minister." Here, as elsewhere, Justi relied on many levels of "evidence" to establish the truth of his cameral-chemical arguments. Not only did he refer to the authority of chemists and insist on his own experimental credentials, but he also used asides to demonstrate that powerful state officials valued his chemical knowledge. This was emphatically not the world of gentlemen witnesses and the Royal Society, not a manufactured public space for collective witnessing.[20] Rather, Justi's digressions always gestured to the secret sphere of the Kammer and the minister's chamber. What happened there remained secret precisely because it was of value. It was here, away from the public sphere, that princes and their ministers decided which projects to fund and which to reject; it was here that contracts were signed, commissions granted, and reputations made. Justi's writings, like those of so many other cameralists, basked in the afterglow of ministerial mystery, promising to offer the public fleeting glimpses of a closed fiscal-juridical world.

Drawing on his predecessors, Stahl and Becher, Justi claimed to have created his own "structure of knowledge about how iron ores and ironstones come into being, and when I have drawn further conclusions and conducted more experiments, it has still never led me astray." He had used magnets to demonstrate that there was no iron in iron ore, but only "an earth that tends to metalness." Phlogiston, lodged in the charcoal, actually created the iron. Even by the standards of his time, when leading chemists were debating basic principles, Justi's claims were a little eccentric and, some might say, just plain confused.[21] Be that as it may, his methodological convictions are worth noting. Justi consistently argued that science—that is, a body of ordered principles that could be taught—could transform the material world, and this applied to fiscal-administrative sciences in just the same way as it applied to natural sciences. He rejected the authority of artisanal practice, claiming that one had to cast off old habits and embrace systematic knowledge, a theme that pervades

Justi's chemical writings, where he regularly railed against the seduction of comfortable routine.[22] In all cases, the same principles applied: whether in chemistry or finance, the improvement of the common good was always to be the guiding aim, a lesson that applied equally to professors of philosophy and ironmasters.

JUSTI'S *SYSTEM DES FINANZWESENS*

Justi's chemical writings had demonstrated his systematic knowledge about steel and ironplate, but another major work, the *System des Finanzwesens,* appeared just as he settled in the Neumark to establish his ironplate and steel works. The *System des Finanzwesens,* typically regarded as a systematic work on public finance, was also a sales pitch, for it was just around this time that he needed Frederick's unconditional support for his venture in the Neumark. Even as district forest officials, provincial finance councillors, and members of the General Directory started to express their doubts about his proposal, Justi dedicated the book to Frederick, his patron, complete with a fawning dedication—the king of Prussia, he oozed, was a happy combination of Julius Caesar and Marcus Aurelius. In addition, Justi acknowledged his intellectual debt to Frederick's *Anti-Machiavel,* which had argued that every sovereign had an interest in promoting the welfare of his people.[23]

Interestingly, though, Justi apologized for some mistakes in the book, explaining that there had not been time to fix them. These were not small mistakes, like missing pages or faulty chapter headings, but went right to the heart of the *System des Finanzwesens.* At the very moment that Frederick was handing over the excise to a group of French tax farmers, Justi had criticized revenue farming in the harshest terms.[24] Despite many years of attacking revenue farmers and the excise, Justi tried to blame his *faux pas* on Montesquieu, who had "painted the system of farming the finances in the blackest colors." According to Justi, now backpedaling fast, this principle applied only to weak princes, who allowed their favorites and public servants to take over the government. Only then, he explained, did revenue farmers become the "bloodsuckers of the people." Justi wrote his dedication (and his apology) in the sleepy town of Landsberg on the Warte River, about fifteen kilometers southwest of his proposed steel and ironplate works along the Zanze. The date was 7 April 1766.[25]

You might say that Justi was a victim of bad timing, for it was just now, in early 1766, that Frederick II instituted one of the greatest fiscal-administrative reforms of the eighteenth century, installing a French-model General Excise

Administration, the *Regie,* in Berlin. The reform came as a shock to many, especially the members of the General Directory and the provincial Kammern, the traditional heart of the Prussian fiscal administration. The establishment of the *Regie* demonstrated, in dramatic fashion, Frederick's boiling frustration with the existing revenue system. On the heels of the Seven Years' War, the king had experimented with new ways to raise revenue—everything from lotteries to state tobacco farms—but nothing seemed to work. The General Directory had resolutely resisted a significant increase in excise taxes, even though that seemed the quickest and most obvious way to raise additional revenue. Frederick, fed up with his ministers, began negotiating seriously with a group of French entrepreneurs to farm the Prussian taxes in January 1766. A few months later, an entirely new, very large, and very French ministry had been created alongside the existing fiscal administration. It represented a direct challenge to the core values of a Prussian state administration that had been conceived and constructed five decades earlier, under Frederick William I.[26]

Given what was going on in Berlin, it was more than a little embarrassing that Justi's new book condemned revenue farming in the harshest terms. It might be the easiest way to collect income for the state, he explained, but it was not the best. Rather, revenue farming along the lines of the *Regie* was "best suited for despotic regimes, and is the true mark of the same."[27] Try as he might to blame it all on Montesquieu and to distance himself from published criticism of French-style tax farmers, it was not convincing. Justi had long criticized this type of fiscal administration. The *System des Finanzwesens* not only attacked revenue farming as despotic, but it was filled with attacks on French-style revenue collectors, who bled the state ruthlessly without any concern for building long-term fiscal resources. "No one," he claimed, "is in my opinion less of a cameralist than the so-called *Plusmacher,* who only seeks to enrich the sovereign, without thinking about the general welfare of state and people." Nor did it take any great skill to increase revenues over the short term. "If it is permitted to oppress the land for the collection of revenue, then one does not need any great Cameral Science." In other words, one needed only force and greed to fill the coffers. More specifically, Justi associated excise taxes, state lotteries and monopolies with *Plusmachen,* which he in turn connected with France, Italy, and all things Machiavellian.[28] The *Plusmacher* was a reckless projector, a false cameralist who catered to every whim and passion of the prince while neglecting the true interests of the state.[29] The virtuous policies of Frederick William I, a model for all "true cameralists," had concentrated on long-term projects, like domain administration and forestry. Now, ironically, Frederick's French tax farmers, the embodiment of bad cameralists, had appeared on the scene just as Justi's book went into print.

The *Finanzwesen,* with its praise of Frederick William I and its censure of revenue farming and excise taxes, could have become a real embarrassment for Justi. But the man knew how to flatter, seduce, and adapt. Though his public writings condemned many of the administrative changes that Frederick was just then introducing, his private memoranda promised pots of money from new mining and agricultural projects. Justi may have excoriated French and Italian fiscal systems, but he was no insular German bumpkin. This was a man of the world, a cultivated French-speaker who had served the Austrian empress and the kings of Denmark and Britain—and he always made sure to present himself that way.[30] In the end, Justi convinced Frederick that he could be a good Prussian *Plusmacher,* a successful projector who would increase the revenue by revolutionizing Prussia's "mining œconomy." In public, Justi unfailingly played the role of the good cameralist, the fiscal official who looked only to the best interest of state and people. Behind closed doors, in the secret councils and memoranda of the Kammer, Justi adopted an altogether different role as an unparalleled projector, pitching one great enterprise after another and promising great profits for king and treasury. In public, he discoursed about the "general welfare" and the "common good"; in secret, he wrote only and always about the "royal interest."

Less than two weeks after dedicating his *System des Finanzwesens* to Frederick II, Justi sent a secret memorandum to the General Directory about the proper management of leaseholds in the royal ironworks. It was just at this time, after arriving in Landsberg, that he was looking to secure favorable terms for his many metallurgical projects in the Neumark. Justi, who hoped to solidify his position as the foremost expert on ironworks in the kingdom, assumed a remarkably condescending tone in the memo, and he made sweeping allegations of systematic mismanagement and fraud. Though he had barely arrived, he was already making enemies.

In Justi's opinion Prussia's "finance collegia," especially the General Directory, had "adopted certain principles that are not only in and of themselves completely mistaken . . . but are also harmful to the highest royal interest." These collegia, Justi complained, had gotten in the habit of releasing tenant-ironmasters from their contractual obligations. Typically, these state entrepreneurs would rent an ironworks from the crown for several years and pay a set quarterly lease over the life of the contract. Lease payments were often based on an estimate of "blow weeks," that is, the number of weeks that a blast furnace would be "blown in" each year. Most contracts called for a blast furnace to be blown in (i.e., operating) for forty weeks each year. Leaseholders, however, frequently petitioned for rent reduction when the blast furnace had not blown for the full forty weeks, and the authorities typically granted such petitions.[31]

According to Justi, Prussian fiscal officials, ignorant of common practice in ironmaking lands such as Sweden and Saxony, had adopted the wrong principles. The policy of rent reduction, he continued, contradicted the "nature" of a tenancy contract, which was based not on the "industriousness or special attentiveness of the leaseholder" but on the "nature of the leased properties and their œconomy." Successful royal ironworks, like successful royal farms, depended on the insight, ingenuity, and industriousness of the leaseholder. But no reasonable fiscal official would try to determine whether a tenant had "made the best use of the weather with his plow and planting, whether he had used skilled workers, whether he had used good seeds, and the like." Yet that is exactly how Prussia's Kammern were treating the royal ironworks, even though "it is just as impossible to determine whether the entrepreneur of an ironworks has done everything necessary as it would be to determine for the leaseholder of a farm."[32]

At the center of this administrative confusion was the blast furnace. Everything depended on the blast furnace, for it was the source of raw iron to supply the forges, fineries, and smithies that together formed the "iron œconomy." When the furnace stopped running, when it blew out, no raw iron could be produced for the ironworks. Unlike the General Directory, however, Justi rejected the notion that royal ironmasters should be forgiven when their blast furnaces blew out. Rather, they should be held accountable, just like tenant farmers. If the blast furnace failed to blow for a full forty weeks, then that was the entrepreneur's problem, and he should be required to pay the rent. Otherwise there would be no accountability, because no investigation could determine why a blast furnace blew out. The workers, "who are all in the pay of the entrepreneur," would never accuse him, or their fellows. Moreover, clever entrepreneurs knew how to sabotage the furnace by sneaking in at night and overcharging it with limestone or ironstone. No one would be able to prove a thing.

But why would leaseholders sabotage their own furnaces? Justi had an answer for that too. The trouble, he explained, was that ignorant officials typically based the entire rent on "blast weeks," as if the blast furnace accounted for most of the profit. In practice, however, it was the forges that yielded more value for each hour of work. In fact, one could produce enough raw iron with the blast furnace in twenty weeks to keep the forges busy all year.

Clever entrepreneurs blew out their blast furnaces after less than forty weeks, petitioned to pay reduced rent, and continued to produce iron wares in the forges and smithies for the rest of the year. In this way, ironmasters increased their own profits while swindling the royal treasury. Justi claimed that it was especially bad in Prussia, where "it has suited the particular interest of the entrepreneurs to blow the blast furnaces not much more than twenty weeks each year." The cast-iron wares were so poorly manufactured that they could not be sold, and no entrepreneur wanted to be burdened with a great stock of useless things. In addition, Prussian ironworks lacked the forge capacity to work all the raw iron that would be produced by the blast furnace in a full forty weeks.[33]

Justi was suggesting a systematic conspiracy to defraud the crown. He claimed that the existing ironmasters—entrepreneurs with leaseholds on royal ironworks—had employed "all secret arts and measures" to justify short blast-furnace campaigns. Relying on precedent and existing practice to secure reductions in rent, they operated their furnaces for only twenty weeks. Seeking only "their own interest," leaseholders shut down their blast furnaces once there was enough raw iron to supply the forges, which continued to operate. This, in turn, caused "the greatest disadvantage to the royal interest." Greed and self-interest, not technical issues like water power, were thus "the only true reason" that Prussian blast furnaces so rarely completed a forty-week campaign. Foreign blast furnaces, by contrast, routinely operated for a full forty weeks, and often longer than that.[34] If the authorities held Prussian ironmasters to their contractual obligations, as in foreign lands, the crown would benefit.[35] Justi's memorandum essentially accused Prussian ironmasters of swindling the king; it also suggested that fiscal officials in the provincial Kammern and the General Directory were incompetent. Even by Justi's abrasive standards, it was quite a performance.

Justi used all the tools at his disposal—public writings about metallurgical chemistry, secret memoranda about crooked ironmasters, essays about good cameralists, private audiences with the king—to convince Frederick about the great untapped revenue potential of the Prussian hinterlands. The old cameralist had an answer for everything. Why did Prussia have to import so much foreign iron? Because Prussian ironmasters were self-interested, dishonest, and ignorant. Could Prussia supply herself with good tinplate, blackplate, and steel? Certainly, but one needed the right officials; skilled cameralists armed with the best chemical and financial principles. But there was one crucial missing ingredient: these magical officials, who knew how to transform Frederick's backward provinces into profitable lands, also had to be honest. But why should he trust them? Why should he trust Justi? The old cameralist had an answer for that too.

HONEST CAMERALISTS

The most prominent section of Justi's *System des Finanzwesens* was not about income and expenditure, population theory, Montesquieu, or excise taxes; it was about the "character and attributes of a true and a false cameralist."[36] Many of his earlier writings, notably *Staatswirthschaft,* had touched on this theme in passing. Now, in 1766, with a lucrative contract in the balance, Justi expanded it and made it pivotal for the whole field of "financial science." True cameralists, he explained, were those who "work for the welfare of states and the true happiness of the people." They had to be distinguished from false cameralists, who ruined the people and, as a direct result, injured the state and its sovereign. By painting a beautiful picture of the true cameralist, Justi explained, he hoped to offer a model for young people; by revealing the ugliness of the false cameralist, on the other hand, he intended to arouse "hate and disgust." The axioms of finance were useless without good cameralists; everything depended on them.

Above all, a true cameralist had to be "an honest and upright man." A perfect state would have no regulations and police ordinances, because all of its members acted like good cameralists, driven by the purest of motives. The people may have been corrupted, but one still needed pure officials, free from self-interest, above gifts and bribes, dedicated only to the true interest of the commonwealth. These good cameralists rejected "private interest," following only the "true interest of the prince and of the whole state." Even more, the good cameralist had to be a "true patriot; that is, he must sincerely love his prince, his state, and the common good." Without such love, he would never be an honest official, dedicated to the common good. "This common good must direct all of his actions, and it must be the fire that swells his veins and warms his principles."[37]

True cameralists understood intuitively that prince and people shared a single, common interest. What harmed the commonwealth harmed the prince, and true cameralists knew of no "fiscal interest separated from the common good." False cameralists, with their Machiavellian tricks, might pursue the "apparent interest" of the crown by collecting revenue in ways harmful to the public. True cameralists found no value in such schemes, "the powerful idols of so many courts and finance chambers (*Finanz-Kammern*)." The true cameralist was not some technocrat, but a *moral* being, temperamentally incapable of harming the public. Though Justi admitted that there were bad cameralists, he considered it unfair that all cameralists were regarded as hard-hearted tools of the prince. On the contrary, true cameralists had to be brave, willing to resist the prince's temptation to pursue his "apparent interest." And the wise sovereign would listen. Frederick William I of Prussia, Justi assured his readers, had been such a king.[38]

Impeccable honesty and integrity constituted the necessary attributes of any good cameralist. But that was only the beginning. The job also demanded vast knowledge, even genius, to separate worthwhile ventures from misguided projects. One needed not only "particular cameralists," who understood some single aspect of the "state œconomy," like the mines, or the forests, or the coinage; one needed "universal cameralists," who comprehended all the different "œconomies" that made up the state finances. The universal cameralist, then, understood the essential principles of everything from mining and forestry to manufactures and chemistry. He comprehended all of these subjects not only individually, but also as a systematic, connected whole.[39] Clearly, then, "no one can be a great cameralist who does not possess great genius." Above all, he would command a great power of invention; he would be a projector of state, able to devise successful projects himself but also able to judge the ventures of others, because the state finances offered "fertile ground" for projectors, who "multiply here of their own accord, like sponges." Any "true and excellent cameralist," had to be capable of distinguishing good state enterprises from bad. Everything depended on that.[40]

The only proper way to raise additional revenue, then, involved creating "new œconomies of state," or expanding the old ones. In other words, one needed to discover new mines, or institute new processes, or unearth new markets. Any idiot could drain more revenue from existing resources, but it took a genius to create new wealth. "Police science" (*Policeywissenschaft*), the carrying card of every universal cameralist, involved the "knowledge and ability to maintain and increase the total wealth of the state." Everything followed from this. It went without saying, of course, that every true cameralist kept meticulous books and observed tireless diligence. "Forgetfulness, mistakes, errors, the usual excuses of disorderly and careless people, have no place in fiscal affairs." Fraud and corruption often hid behind such excuses.[41]

Justi's "bad cameralist" was the opposite of the good one. He might not murder or break into houses at night, but he was not good. The bad cameralist worked to advance his own interests by pleasing the prince. "The will of the prince, the satisfaction of the prince's passions, and the prince's apparent interest"—these were the only things that drove him. Accordingly, he would never contradict his lord, seeking only to justify the sovereign's actions even as he impoverished the people. Above all, bad cameralists prayed to a false god called the "fiscal interest, or apparent interest, of the prince." Europe's "finance collegia" were filled with bad cameralists, raised in the daily business of the Kammer and "without any rigorous sciences," who prayed to this idol. Corrupted by bad principles, these officials mistakenly believed that the "profit of the prince is the single mainspring of all state finance."[42]

The surest mark of the bad cameralist, and the surest symptom of a corrupted state, was an explosion of excise taxes. Bad cameralists had found ways to tax everything, even the necessaries of life, and to raise revenue in all the wrong places. "Now they increase the taxes on immovable goods, now on consumption, now on tolls, now they demand the tenth penny, and the twentieth, and the hundredth, . . . now they lease out all the state revenues and allow the leaseholders to pay them great advances, now they permit inferior coin to be minted." But such techniques demanded no special knowledge or wisdom. Every "village schoolboy" could employ them, and yet "it is just these inventions that are seen as the great accomplishments and skills of the cameralists, which one believes must be rewarded with honors, titles and riches."[43]

For Justi, true cameralists had to be forged in the crucible of systematic, scientific knowledge. The important thing here was not applied science, the utility of the latest chemical theory, or the system of finance; the important thing was character. The mastery of a science and devotion to its first principles provided the kind of detachment necessary for fiscal officials, who were every day confronted with the temptations of office. Only such detachment—a kind of proto-objectivity—could ensure honest and principled administration of the crown's income and expenditure. As a textbook writer, academic and mining official, Justi himself bore all the marks of the good cameralist.

THE BERGHAUPTMANN COMES TO VIETZ

Once Justi had secured a commitment from Frederick, he began touring Prussian ironworks. At the same time, he began preparing the groundwork for his personal iron empire in the Neumark by complaining about almost everything. There was, for example, trouble about the lease projections. Factor Helmkampf, a provincial official, had begun making revised estimates about the correct amount of annual rent for the ironworks in Vietz and Kutzdorf. Justi, however, would have none of it. In his opinion, Helmkampf had no authority to revise the estimates. "I do not see," he complained to Berlin, "how these new estimates by Factor Helmkampf could be in keeping with the highest royal purpose." Nor did he think that Helmkampf had authorization from the General Directory to make such estimates. The General Directory, not quite knowing what to do with Justi, forwarded the letter to the provincial Kammer in Küstrin. Marginal notes on the correspondence indicate that there was already confusion about Justi's position. Should he be treated as a potential tenant-entrepreneur, reporting to the provincial Kammer, or should he be treated as a royal Berghauptmann, with direct access to the king and General Directory?[44]

In the Neumark, there was concern about water and wood. Some provincial officials doubted whether there was enough water to power another blast furnace, but Justi rejected their concerns.[45] Only during the summer months would there be any lack of water, he argued, and the new furnace was necessary, because the success of his Zanze works demanded an adequate supply of good iron, as the king himself agreed.[46] Access to wood for fuel provided another source of constant conflict. The blast furnace demanded large amounts of charcoal, which in turn demanded considerable quantities of wood from the surrounding forests. When Justi requested twice as much wood as the previous tenant, Winckelmann, local forest officials started to complain.[47] Justi explained that he needed additional wood to fuel a second blast furnace in Vietz. Only in this way could he satisfy the increased demand for iron that would be occasioned by the new works on the Zanze.

Provincial forest officials were not convinced. Forstmeister Sohr in Küstrin, for example, worried that Justi's expanded iron and steel operation would devastate the forests of the Massin district. Justi responded that there was nothing to worry about, "because I intend to introduce better wood management there." He thus "directed" the General Directory to assign him four hundred *Klafter* of wood and to issue the necessary orders to the Kammer in Küstrin.[48] "I will also seek to introduce wood management in the operation of both blast furnaces," he wrote the king, "so that I will never request more than half the amount of wood for the second blast furnace as is required for the first." Justi did not stop there, but went on to level indirect criticism at the administration of Forstmeister Sohr, arguing that all of these worries about wood could have been avoided through better forest management. One only had to focus more on "conservation and replanting of the forests, which until now has been very absent in the Neumark."[49]

The familiar legend of a blind, old, incompetent Justi groping his way through the Neumark is in need of serious revision. During April and May of 1766, he was a formidable force, threatening locals, demanding detailed administrative records, publishing large books, traveling throughout the kingdom, and assuming special privileges for himself all over the place. At the beginning of his tenure as Berghauptmann, Berlin gave him almost everything he wanted, much to the chagrin of local and provincial authorities. If the new combined royal ironworks were to succeed, Justi claimed, he would need another blast furnace, housing for additional workers, and a greater wood allowance. In addition, contrary to the claims of Factor Helmkampf, he insisted that the ironworks had seen little improvement for many years and were in need of new investment.[50] By the end of April, he had secured permission to build a new house in Vietz for the workers. The project would be bankrolled by the king, and the Kammer would provide the necessary wood, free of charge. But even as he arranged to take over the ironworks in the Neumark, Justi also had separate duties as a Berghauptmann, traveling to various royal ironworks, especially the complex in Torgelow, where he had just "perused 14 thick volumes of records."[51]

At the end of April, Justi demanded that the Kammer in Küstrin send him complete records from Vietz and Kutzdorf. He wanted to examine all the inventories and lease contracts before the official transfer of the ironworks at the end of May. When officials in Küstrin refused to send Justi the records, things turned ugly. "Since his royal majesty has graciously entrusted me with the direction of all mines and ironworks in his lands," he blustered, "the communication of records that deal with mines and ironworks must naturally be implied in that charge." No other Kammer, he continued, not even the General Directory, had rejected such requests, and he threatened to get a cabinet order to force Küstrin's hand. Ultimately, the Kammer did send along the records, and a few days later Berlin directed the Neumark to provide Justi with all the wood that he needed, free of cost, from the forests of the Massin district.[52] With the official transfer of the ironworks less than a month away, Justi was making sure that his lease would have the best terms, and the authorities in Berlin were giving him everything he wanted.

THE TRANSFER

On the morning of 31 May 1766, Justi gathered in Vietz with members of the provincial government to effect the formal transfer of the ironworks. With Justi were the previous leaseholder (Winckelmann), the president of the Kammer (Pappritz), a war and domains councillor (Jaeckel), the chief smelting factor (Helmkampf), the controller (Wagner), and the scribe (Schöning). The ironworks lay about twenty kilometers east of Küstrin, near Zorndorf, where Frederick II had defeated the Russians only eight years earlier. Kutzdorf, no more than a mile or two from the heart of battle, had been completely flattened, and the forests of Massin set ablaze.[53] The Russians may have lost that battle, but they successfully ravaged the area, burning whole villages and forests as part of the operation. Justi's new home, depopulated and deforested, still bore all the scars of that conflict. It must have been grim.

Justi toured the blast furnace and foundry in Vietz together with the previous leaseholder, Winckelmann, and the official witnesses. There were frequent debates and arguments along the way, as the parties haggled over charcoal stocks, wood, artillery casts, raw iron, and all manner of implements. The process continued for two full days, moving on to the hammer forges and blacksmith shops in Kutzdorf on 2 June. At the end of that second day, after weighing the raw iron in front of witnesses—there were 1,119 centner on hand—Justi handed Winckelmann a credit slip for three thousand taler. Then he left abruptly for Stettin on official business.[54]

Leasehold contracts were the bread and butter of provincial administration. The inventory books had to be checked and double-checked against the condition of the property; the raw materials had to be weighed and assessed; the accounts had to be verified. It was a painstaking business, dedicated to specifying every groschen of added or subtracted value. Every window pane, every door jam, every lock and fencepost had to be accounted for.[55] Justi was quite good at it, dealing and bargaining as he went. But the only real cameralists there that day were Pappritz and Jaeckel, representatives of the Kammer in Küstrin. Justi's many books were designed to train officials like them, who leased out royal domains and enterprises to the highest bidder. In theory, these servants of the public existed over and against private leaseholders; in practice, important leaseholders almost always had official titles and quasi-official roles. Justi the good cameralist saw a conflict of interest in such hybrid functions; Justi the leaseholding state official did not complain about it.

Of course, it was unrealistic to imagine that an important *Pächter* like Justi could function as a merely private individual, for when he moved to Vietz in the summer of 1766, he acquired not only foundries and blast furnaces, but people too. From the day he arrived, Justi began to behave like the governor of the whole district. His leaseholdings now included charcoal magazines, two hammer forges, a foundry, several smithies, a brewery and distillery, a stable, several family houses, a factory building, and thousands of pounds of iron and iron ore. Equally important, however, were the iron workers that he inherited (they were inventoried as well): furnace masters, casting masters, smelters, charcoalers, and slag haulers. The skilled workers in Vietz and Kutzdorf were almost all "foreigners" like Justi, from places like the Harz Mountains, Silesia, and Saxony.[56]

Foreign workers, however, brought more than skill to the Neumark, as Justi soon discovered when he became embroiled in a struggle with two weavers—he called them "Matthes and Kindel"—who were living in the manufacturing house with their wives and daughters. Justi had arranged to have foreign metal workers brought in, but in June, when they arrived in Vietz, they had nowhere to live. He ordered the weavers to leave, but they refused. On 16 June, he filed an official complaint with the district office, explaining that "already on the 2nd of June they were told to leave their apartments in the manufactory within 14 days." The weavers, he explained curtly, were blocking efforts to manufacture blackplate at the hammer forge in Kutzdorf, because the new workers had nowhere to live. "The ironplate and steel smiths cannot simply lie on the street, but must be afforded apartments."[57] But the weavers paid no attention.

At this point, Justi assumed a quasi-judicial and police role, reminiscent of his efforts to arrest beggars in Göttingen a decade earlier.[58] He gave the weaver Matthes twenty-four hours to clear out and threatened him with punishment. But "weaver Matthes replied that he would not leave the manufactory house, and he has not made the least preparation to do so, despite the fact that it is noon and the 24-hour deadline has now passed." Justi was outraged. He now "required" the district office to send "a court servant along with several men" out to Kutzdorf "this afternoon" and "have the weaver Matthes's things, along with himself, evicted." The other weaver, Kinell, would have three more days to get out. Justi wanted speedy justice, and he warned the district *Amtmann*, Lehmann, that any failure to act would injure "the royal interest" because the whole operation was being financed by the crown. Amtmann Lehmann was clearly baffled by Justi's letter. If the Berghauptmann's orders had been ignored, why would they listen to him? And if Justi had not managed to evict the two weavers, how was Lehmann supposed to manage it? He could not just show up that afternoon and throw them into the street. He counseled Justi to seek authorization and help from the Kammer in Küstrin.[59]

As the summer wore on, Justi's troubles multiplied. Councillor Jaeckel, a member of the Kammer in Küstrin, started to raise serious questions about how Justi was spending the king's money. The crown had provided a substantial fund to build the new works on the Zanze, but Justi was drawing on that fund to settle his accounts and pay his lease in Vietz and Kutzdorf. Jaeckel considered this a misappropriation of royal funds, and he refused to countersign some of Justi's credit slips at the provincial treasury, blocking payment and causing the Berghauptmann considerable embarrassment.[60]

Justi tried to explain the problem away, but by mid-July Jaeckel's concerns had become a real issue. Was Justi, as Jaeckel suspected, covering his operating expenses in Vietz and Kutzdorf with funds designated for the *new* royal works on the Zanze? Justi told the authorities that "in this first period, during which these ironworks have almost no hope to sell enough," he would need to draw on the Zanze fund to pay his rent. Since there was a special fund to purchase iron for the Zanze works, Justi did not see any problem. Since the ironworks in Vietz and Kutzdorf would supply the Zanze works with iron, it was all the same. From the perspective of the Kammer, however, this was an outrageous claim. After all, the previous entrepreneur, Winkelmann, had managed to pay the rent, so why couldn't Justi manage it?[61]

Even as provincial councillors began to question his administration, Justi pushed aggressively forward. By mid-July he was reporting great success with his ironplate experiments in Kutzdorf. The tests, he claimed, were working well, both with local and Swedish iron. If this were true, Kutzdorf would soon be ready to supply the Zanze works with ironplate. In that case, however, Kutzdorf would need more raw iron, which meant that the second blast furnace in Vietz had to be blown in. But the new blast furnace workers had to live somewhere, so Justi requested five hundred taler for a new family house in Vietz. Despite some misgivings, the General Directory ordered Küstrin to give Justi what he needed. He had convinced them that the operation needed more raw iron.[62]

After Justi secured approval for the new family house, Vietz experienced a small construction boom. In addition to the family house, the second blast furnace needed a lot of work; its water supply had to be extended, it needed a new water wheel, and the bellows arch needed fixing. Even more ominously, the amount of wood for charcoal burning would double. Justi asked for permission to open a brewery and distillery in Vietz for the workers. The wood for all of these projects was valued at more than two hundred taler. All of it would be provided free of charge from the forests of Massin. Construction costs would be borne by the crown's Zanze fund.[63] By the end of September, Justi was still using the Zanze funds to pay his quarterly rent.[64]

Even as he made plans to expand the ironworks and to bring in new people, Justi suspected that his own workers in Vietz were defrauding him. More specifically, he suspected that the furnace masters and the foundry workers, scheming together, were scattering raw iron among the slags, and then recovering it later under false pretenses. Since it was impossible, "even with the greatest watchfulness," to stop such a swindle, Justi suggested that all Prussian ironworks stop washing and pounding the slags to recover scrap iron; the returns were minimal, and the risk of fraud was too great.[65]

THE WELL-POLICED FOREST?

While he haggled with the provincial Kammer over financing and struggled with the perceived duplicity of his own ironworkers, Justi had other problems closer to home, in the royal forests around Vietz and Kutzdorf. His many construction projects had already taken a toll on local forests. Now, with the prospect of a second blast furnace, the demand for wood was expected to double.[66] He had clashed with local foresters since the summer, and now, in November, these local clashes erupted into an official dispute that ultimately had to be mediated by the General Directory and the king.

The quarrel centered around the fir forests and heath of Massin, where district foresters had taken issue with Justi's approach to harvesting wood. This area had witnessed the battle of Zorndorf not long before, when Frederick and his artillery clashed with Russian troops in the summer of 1758. Thousands had died, horses and men had lain trampled and bleeding in the boggy quagmires. Vestiges of the battle still scarred the landscape. In the fir forests and marshlands of Massin, there were still burned trees all around. Zorndorf, burned to the ground by the Russians, still bore the marks of that conflagration. In Kutzdorf, where Russian cavalry fled the battle, hammer forges and blacksmith workshops had been completely destroyed. And in Vietz, the Russians had burned everything—blast furnaces, houses, and charcoal sheds.[67]

Oberjäger Kleyensteuber, who supervised the forests around Massin, criticized Justi for irresponsible management. According to Kleyensteuber, Justi's wood gatherers were harvesting wood arbitrarily and without any direction. The principles of good forestry dictated that they should be gathering deadwood; in practice, however, they were out in the heaths and forests cutting down everything. Justi was ignoring all good rules of "conservation." Even worse, his inattention to good forest principles was undermining the long-term viability of the ironworks, because all the available fuel wood would be gone within a few years. Kleyensteuber insisted that Justi's furnacemen, charcoal burners, and wood gatherers had to be controlled before it was too late.[68] Justi brushed these complaints aside. After all, who knew more than he did about the principles of good forestry?

Justi had conjured beautiful visions of well-policed fiscal forests in several works, most recently in his *Finanzwesen*.[69] Echoing the work of other cameralist mining and forest officials, he made a strong case for replanting, sustainability, and conservation.[70] In a well-ordered state, he explained, the Kammer should have maps and descriptions of every forest district, with detailed information about tree varieties, soil types, wood stocks, streams and rivers, denuded areas, and the location of all mining and smelting operations, so that cameralists could survey the sovereign's forests at a glance. This would allow them to "draw as much income as possible" from the woods without harming them. It followed that there could be no completely barren spaces in the forest. Even the driest, hardest, sandiest ground could support some tree varieties, so there was no excuse for empty spaces, which appeared "when wood cuts happen at the wrong time, when forest officials do not watch the wood gatherers and allow them . . . to cut down live trees." For Justi, denuded forests provided physical evidence of negligent administration, a clear sign that the prince's foresters and cameralists had failed to do their duty. Healthy forests, by contrast, provided the surest sign of good government. Every good cameralist was committed passionately to conservation and reforestation. "If he has other inclinations and attitudes, one can certainly expect little diligence from him in all other affairs."[71]

And if the state's forests were less than perfect, Justi had the solution. No expense should be spared. In every district, denuded areas could be used for vast nurseries; these nurseries, in turn, would be used to reforest empty spaces. Members of the Kammer would take the lead, employing the latest botanical and agricultural knowledge. He did not stop there, but showed exactly how it should be done. One had to dig ditches at regular intervals, fill them with foliage and brush, and allow it to compost. This would provide good soil, "so that in every hole, where a seedling will be planted, a couple of shovels-full of this earth can be put in." The Kammer should even extend its vigilance to irrigation, making sure to have the seedlings watered in case of drought. "The water can be taken out of the nearest rivers, streams, and lakes, put in a large barrel, and brought up on a wagon with a horse."[72] Even at a distance of more than two centuries, you can feel the seductive power of Justi's rhetoric, which combined the integrity of the idealized fiscal official with the wisdom of practical experience and the promise of science. The man had a good ear for the power of small, specific, everyday details. He always sounded as though he was writing from local experience, even as he benefited from the cosmopolitan knowledge of the well-traveled savant. He was an unparalleled self-promoter.

In Vietz, however, Oberjäger Kleyensteuber had developed a somewhat different impression of Justi's administrative gifts. "My duty and conscience do not allow me to watch with indifferent eyes," he wrote Frederick, "as these people hack down standing wood foot-by-foot, while the rich deadwood, which is also the best for charcoal, lies around and spoils." Kleyensteuber had insisted repeatedly that Justi's wood gatherers should collect only fallen deadwood for charcoal burning, but the Berghauptmann had ignored him completely.[73] "The smeltery cannot last more than a few years at this rate," Kleyensteuber continued. Justi responded by adducing any number of important "forest principles" to defend his actions.[74]

Berlin ordered the Kammer in Küstrin to look into the dispute between Kleyensteuber and Justi, and on 17 November the investigating commission issued its report, which not only rejected all of Justi's claims, but also expressed dismay at his behavior. "We should neglect our duty," the commissioners wrote, "by allowing him to exercise unchecked and arbitrary management in his royal majesty's forests." Kleyensteuber's plan, which would force Justi to get his deadwood from three specified locations, might incur slightly higher transport costs. But the alternative meant "the ruin of the forest" as "hundreds and even thousands of cords of deadwood" rotted there. The commissioners seemed amazed by Justi's presumption.[75] "This transport of charcoal, which seems so burdensome to the Berghauptmann, is apparently the basis of his whole complaint." Justi, who was carting coal at his own expense, seemed willing to sacrifice the royal forests to save himself some money. The commissioners, joined at the end by *Oberforstmeister* Sohr from Küstrin, suggested that it was time for Justi to "calm himself."[76]

Despite Justi's efforts to sway Frederick with impressive flourishes about the principles of good forestry, the king was clearly annoyed. His personal response to Justi verged on the sarcastic. "Whether it is completely in keeping with the rules of good forestry, when, in the heaths, whole forest districts are cleared of wood all at once, and after subsequent replanting and reseeding are cleared again; still, such rules cannot be practiced in the forest districts of the Neumark, especially in those which, because of remoteness and lack of consumption, are filled with much deadwood."[77] Berlin sided with the Kammer and decided to enforce its recommendations. Justi's wood gatherers would be restricted to collecting deadwood from locations specifically designated by the king's forest officials. The letter ended with a harsh rebuke, emphasizing the limits of Justi's authority. No one was permitted to fell standing wood, especially not healthy trees, without explicit permission from a forester. This rule was intended to protect the forests from harm. In future, "arbitrary" harvesting by Justi's charcoal burners and woodcutters would not be tolerated. The letter scolded Justi for attacking Kleyensteuber, who had simply been doing his job, and it turned his well-known "forest principles" against him. "Because the Berghauptmann himself acknowledges that green wood is not as effective for charcoaling as dry wood, and that clearing the forest of fallen deadwood is an extremely desirable thing, we cannot imagine, given his recognized insight and aforesaid declarations, that he would really request permission to charcoal-burn green wood when enough deadwood is still on hand."[78] From now on, the ironworks would make charcoal only with fallen deadwood from specified locations. Justi and his workers would receive direction from forest officials. Kleyensteuber had won.

FALL FROM GRACE

Justi's losing battle with the forest administration spelled trouble. In the months that followed he quickly fell out of favor with the king and his ministers.[79] While fighting with foresters in the Neumark, Justi had also managed to give the king bad advice about a mining venture, convincing him to support a cobalt project in Silesia. Frederick was so taken with the idea that he instructed his powerful Silesian minister, Ernst Wilhelm von Schlabrendorff, to give Justi whatever he needed. "Because of circumstances with his vision and because of the work that has to be completed before winter at the works on the Zanze," Justi could not make the trip to Silesia himself. However, the king instructed him to choose another "solid mining expert" who could go in his place. If, as Justi prophesied, there were rich cobalt veins in those Silesian hills, then Frederick promised to make funds available for their exploitation and preparation.[80] As it happened, Justi helped convince Frederick to invest his funds in a fraudulent venture. Silesian linen weavers used tons of cobalt every year to dye their wares. The cobalt had to be imported from Saxony, which cost tens of thousands of taler annually. Frederick was therefore very keen to find another source of cobalt in his lands.

In May 1766 a projector named Herzer informed Schlabrendorff that he had discovered cobalt ore in Rudelstadt, near Kupferberg; to prove it, he sent samples to the Kammer in Breslau, capital of Silesia. Herzer, it turned out later, had gotten the ore in Saxony and smuggled it into Silesia, but in the months that followed he managed to convince Schlabrendorff, and then Frederick himself, that there were rich cobalt deposits in the Kupferberg mines. When the king provided twenty-five hundred taler for exploration and development, Herzer disappeared with the money. Months later, after careful study by mining experts, it became clear that cobalt ore was not native to Kupferberg. Herzer had swindled the king. One can imagine Frederick's displeasure with his own Berghauptmann, who had written so much about the importance of vetting state enterprises with advice from good cameralists.[81]

Justi continued to make proposals and suggestions about the Prussian "mining œconomy" to the king and the General Directory, but the sting of the Silesian cobalt sham stuck with Frederick. By February 1767, he was clearly fed up with Justi's many projects and promises. The king now wanted evidence of real progress at the steel and tinplate works in the Neumark. Justi continued to dodge direct inquiries about the financial situation there, but Frederick had clearly had enough: "I find it very sweet that you are thinking ahead about how to improve mining in my lands, but the main thing is that you find skilled mining experts [*Bergverständige*] and that you engage them; then I will not once again, as happened last year because of the cobalt exploration that you initiated, be subject to completely useless expenses."[82] The king, still annoyed about being swindled in Silesia, was now becoming skeptical about Justi's other claims. Even as Justi offered constant assurances about the progress of the new royal tinplate and steelworks along the Zanze, Frederick was becoming increasingly impatient. The plan that they had developed together, the king explained, "involved much more than the mere construction of buildings." It was time, he insisted, to see some real results. Justi, he continued, seemed to believe that it was enough to report progress about the buildings, as if it were acceptable that everything was half-built. But where were the workers? Where was the steel and tinplate? If Justi expected to produce the profits that he had promised, he would need to "bring in many workers," and he would have to make arrangements for "both domestic and foreign sales of plate, steel and steel wares, something which you do not seem to have considered at all until now." Frederick declared that he would consider no new recommendations and projects until Justi could demonstrate material success in the Neumark.[83]

Correspondence from Frederick's cabinet became increasingly brusque after this. In June 1767, the king ordered Justi to submit a complete budget plan for the coming year, "so that I can see to what extent you will be able to fulfill the revenue projections of 27 July 1765." He assumed that the new works were now in full operation, "because I have not heard otherwise."[84] In response, Justi promised the world. What else could he do? He told the king that everything was in working order, and advised him to place an import duty of 30 percent on all foreign blackplate and tinplate. The king then informed the General Directory that the works on the Zanze had progressed far enough "so that all provinces can be completely supplied with steel and plate," and he ordered his ministers to implement the import duty. But Frederick also issued very specific orders to Justi, instructing him to provide the necessary quantities of steel, tinplate, and blackplate, and to make sure that each province had made the necessary arrangements to supply the public. Justi petitioned Frederick for some relief from the debts he had incurred in Vietz and Kutzdorf, where he had borrowed from the Zanze funds to cover his quarterly rents. But the king was unforgiving. "I will not get involved in that," Frederick explained. "You not only agreed to the existing lease amount when you took over the works, but you also committed to a much higher yield based on the introduction of improved management and manufacturing practices." It was time, he threatened, for Justi to stop delaying; it was time to sell some stuff.[85]

When Justi proved unable to repay the outstanding rent, Berlin initiated an investigation. In September 1767, Justi was relieved of his commission. The letter to Justi from Berlin was direct and scathing. The king, it began, had learned about Justi's "unscrupulous administration" with the "greatest displeasure." Justi's commission as a royal administrator and Berghauptmann was over, effective immediately. Berlin provided specific instructions about records as well. The investigating commission would need to secure all inventories, account books, and receipts; separate reports would need to be prepared about the all the money that was paid out from the Zanze royal fund as well as the unpaid rents in Vietz and Kutzdorf. Justi was warned in the "strictest and harshest terms" not to leave Vietz and to avoid all contact with the ironworkers there. Councillor Schönwaldt would assume responsibility for the operation while Councillor Jaeckel conducted a full investigation into what had happened.[86]

THE AFTERMATH

After Justi was jailed in February 1768, several officials were brought in to salvage the ironworks, but things were even worse than anticipated. The raw iron was brittle. The wood supply had been devastated. The iron ore was miserable. The workers were undisciplined. The account books were a mess. It was a cameralist's worst nightmare. Determined to discover what had gone wrong, Frederick had his most trusted minister, Ludwig von Hagen, initiate an investigation. Hagen sent Secret Finance Councillor Reichard to Vietz, Kutzdorf, Zanzhausen and Zanzthal. The decision to send Reichard, who had served as Hagen's lieutenant in Cleves and now held an important position in the General Directory, suggested the importance of the investigation.[87] Reichard left for the Neumark on 10 June 1768 and submitted his full report ten weeks later, on 26 August.[88]

Reichard had seen tough times in Cleves, but even he was overwhelmed by what he discovered in the Neumark. There were serious problems with budgets and bookkeeping, so that he had to begin from scratch, creating estimates about everything from wood consumption to raw iron production. "The outgoing wares and the incoming moneys spent on them," he wrote, "have been jumbled." Reichard scoured records going back to the 1750s; he interviewed all the workers; he assessed the physical and geographical characteristics of the region; and he interrogated key officials. It was an elaborate investigation.[89]

The report was damning. In Vietz and Kutzdorf, things had deteriorated under Justi, and his new enterprise on the Zanze had been an unmitigated disaster. The steel-refining forge installed by Justi had been so completely useless that it was later converted to a tilt-hammer forge for making bar iron, but even this had generated "more losses than profits." The works in Zanzhausen also suffered from a chronic lack of water, as Justi's critics had initially warned, and the forges sat idle much of the time for lack of a power source. Even worse, there was serious trouble with the iron ore. Justi had assured Berlin that, with the application of proper chemical principles, it could be used to produce the finest iron. He had been wrong. None of the iron ore in the area seemed acceptable. In addition, there were piles of useless cast-iron wares lying all around the works, and many of the ironworkers brought in by Justi were completely "useless and untrustworthy."[90]

In Vietz, where Justi had rebuilt a second blast furnace during 1766, things were worse than they had been under the previous tenant, Winckelmann. Due to a chronic lack of water, the second furnace operated for only twenty weeks each year, so that the two blast furnaces, taken together, operated for a total of no more than sixty weeks annually. Recall Justi's scolding memo to the General Directory about the need to enforce the contractual obligations of its ironmaster-entrepreneurs—the irony was certainly not lost on Frederick and his ministers. In the end, after significant investment by the crown, not much had changed. There was another blast furnace, and there were new buildings, but the place still produced mainly crude, cast-iron shot and a variety of other simple cast-iron wares.[91]

Despite all of Justi's "improvements," the quality of the cast iron was worse than ever, lease payments had dwindled, and the forests around Vietz had been devastated. Winckelmann and Helmkampf, the leaseholders before Justi, had done pretty well, paying the annual rent and producing decent iron for the kingdom. Justi, on the other hand, had paid only 1,843 taler in rent, or less than half of what he owed, while running the operation into the ground. Worst of all, though, he had damaged the reputation of Kutzdorf iron. It would take a long time to repair that reputation and to rebuild markets. Even now, in 1768, more than twelve tons of Justi's bar iron was sitting in a Magdeburg depot, "but until now not the least of it has been sold, and such a sale is unlikely to happen because of its proximity to the Harz." The Harz produced good bar iron, so that the inferior Kutzdorf goods were likely to yield nothing more than embarrassment. Reichard recommended that the Kammer ship Justi's bad iron from Magdeburg back to Brandenburg, where it could be reworked. This, he hoped, would "keep the public from complaining."[92]

As bad as things had gotten in Vietz and Kutzdorf, Reichard discovered the real disaster in Zanzthal and Zanzhausen, where Justi had tried to establish his steel and tinplate works. The promise of domestic "cemented" steel had been especially enticing to the king. Cemented steel, or blister steel, had been around since the seventeenth century, and most of it was made in England, but it was not easy to replicate. One packed the iron in "cement powder"—a mix of charcoal, ashes, and mineral salts—and baked it for a week or more until it absorbed enough carbon to be "converted" to steel. Recognizing the importance of good bar iron for the cementation process, English steel manufacturers had long used Swedish iron, and some of them even insisted on certain brands.[93]

While England produced blister steel in ever larger quantities, the French and Germans tried to figure out how it worked, and they produced a large body of scientific literature in the process. Stahl and Réaumur, for example, were important in shaping Justi's views, which were thoroughly phlogistonist. "Steel," he wrote, "is nothing other than iron whose metallic earth is combined with enough phlogiston that it has disproportionately more hardness than common iron." Once again, Justi mobilized the latest scientific knowledge in the service of his projects, using it to convince the king that he knew what he was doing. Three years later, and with the benefit of hindsight, Reichard reasoned that Justi's "ill-fated blister steel" enterprise never had a chance.[94]

It is hard not to marvel at Justi's confidence. Many ambitious entrepreneurs had failed to reproduce the coveted tinplate of Saxony and Wales, and many more had been unable to mimic the blister steel of the West Midlands.[95] Nevertheless, he not only proposed to produce steel *and* tinplate, he promised to do it with local iron from the Neumark. It is amazing that anyone believed him. Had he proposed such insanity to the Kammer of a mining state, like Saxony or Hannover, they probably would have laughed him out of the chamber. They knew, as Frederick and his officials did not, that the production of high-quality steel and tinplate depended above all on material and local factors—long experience, skilled workers, and the best iron ore. Justi promised the king that science could replace those things, but he was wrong.[96]

In fact, the Zanze works never even got off the ground, because there was never enough water to drive the machines. Those idle machines—the blackplate hammers, the tinplating works, finery forges for the steel—had been expensive to build. Frederick committed a staggering 140,000 taler to the project, and by 1768 the Kammer in Küstrin had already paid out 138,000 taler. "Most of it was spent on the buildings, in completely unnecessary fashion, which ate up half of the funds." The rest had been squandered on foreign workers, operating costs, and many cash advances to Vietz and Kutzdorf to cover the rent. Two years later, after so much investment, it was clear that the Zanze works were a complete failure. Far from producing the increased yields and profits that Justi had promised, the works could not even pay for themselves. Moreover, the records and account books were so disorganized that Reichard could not determine how all the money had been spent. Some thirty-five thousand taler remained wholly unaccounted for, and there were irregularities involving the brewery and distillery.[97]

Reichard's report signaled that it was time for drastic measures. Justi's grand plans for cemented steel had to be abandoned. With any luck, the expensive new works along the Zanze might be converted to produce more pedestrian products, like blackplate and bar iron. The crown could also provide water power by connecting the ironworks to some lakes farther up above with a canal, but that would cost another 259 taler. Berlin directed the provincial Kammer to devote all possible effort to the discovery of decent ore deposits. The many useless iron wares might be melted down, so that the iron could be reused, but this would result in major losses. Reichard suggested auctioning them off in bulk.[98] In addition, many of Justi's workers had to be dismissed, and an example would be made of several forgemen in Zanzhausen, "because these people are worth nothing, and because some order has to be introduced among the ironworkers there." Until better ore could be found, the forges would be provided with Siberian iron, instead of the more expensive Swedish variety.[99] Finally, Berlin stressed the need for better accounts and records. Justi's administration had been out of control, rife with corruption and mismanagement. From now on, there would be better oversight.[100]

About six weeks later, in October, Secret Finance Councillor Ernst arrived in the Neumark to review the situation and implement Reichard's recommendations. Ernst was one of the experienced mining officials brought to Berlin by Hagen in 1768.[101] Now, he had come to the Neumark to clean up Justi's mess. When Ernst arrived in Vietz, Justi's nemesis, Councillor Jaeckel, offered an extensive set of recommendations. In November, Jaeckel submitted a memorandum based on these recommendations to Berlin.[102]

With Justi imprisoned in Küstrin, Jaeckel had taken over the combined works in the Neumark. By November, he had begun to sound like Justi, arguing that he needed new bookkeepers, scribes, and overseers so that he would be free to "manage everything in its completely essential totality." More than technical knowledge or better ore, Jaeckel wanted dependable and reliable people who knew the region and understood the operation. He imagined a system of mutual control, in which "every official can be monitored by the others." This, he continued, would conform with the structure of foreign ironworks, "where every smelting official monitors the others." Under Justi's autocratic management, there had been no accountability. Now, it was crucial to reintroduce the collegial principle, so that mining officials would "control each other" by deciding matters "in council." If these principles were followed, Jaeckel promised, "good order" could finally be established at the combined iron and tinplate works.[103] Ironically, Jaeckel's principles, which every aspiring professional cameralist learned at university, had been articulated by Justi in one textbook after the other.

In the aftermath of Justi's failed administration, Frederick founded the Mining and Smelting Department as the seventh section of the General Directory, and he put his most trusted official, Ludwig von Hagen, in charge of it. Hagen, "the greatest official to emerge in Prussia during the later half of the reign of Frederick the Great," had already assumed a leading position in the General Directory when Justi took up his commission as Berghauptmann in 1765.[104] After Hagen's death, Friedrich Anton von Heynitz, founder of the Bergakademie Freiberg, would take it over.[105] Hagen and Heynitz endeavored to harness science in the service of the treasury, and it has been suggested that they shared "certain characteristic goals of the later cameralists such as Justi," whose ideas influenced their administrative practices.[106] In the end, though, Justi's spectacular failure in the Neumark probably had more impact on the Prussian mining administration than his writings did.

EPILOGUE

The question of Justi's guilt or innocence has preoccupied commentators since the eighteenth century. The consensus now seems to be that Justi was a victim of circumstance.[107] Old, blind, and persecuted by local officials, he was apparently overwhelmed by his duties. I do not believe it. The Justi we discover in the provincial archive was relentless. He petitioned. He schemed. He threatened. He cajoled. Blindness may have overwhelmed him in prison—its creeping effects are visible during 1766 and 1767, as his handwriting gradually deteriorates—but not before that. Between 1765 and 1767, he was constantly on the road, touring Prussian ironworks, and poring over records. In 1766, during one of his many disputes with Küstrin, he insisted on viewing all the contracts and tenancy records himself, refusing to let anyone else do it for him; and before the transfer of the ironworks in Vietz and Kutzdorf, he perused hundreds of pages of documents in a day or two.

But if Justi was not a victim, what was he? Certainly, he made enemies, but he did that everywhere. Justi's allies in Berlin, who were much more powerful than his enemies, wanted him to succeed. For more than a year, they gave him everything he requested. It is tempting to blame Justi's failure on hubris, the tragic flaw of an otherwise gifted cameralist. But that would be too easy. Frederick and his ministers were willing to accept Justi's arrogance, as long as he produced results. Of course, it is possible that those pesky provincial officials were right all along. Maybe Justi really did mismanage things from the start. Maybe he really was defrauding the crown. That, apparently, is what the Prussian Mining and Smelting Department finally concluded, three decades after the fact.

In a letter dated 10 January 1800, the Mining and Smelting Department requested information from the provincial archive. "Frankly, the issue is this: it is well known that von Justi resettled many ironplate workers and ouvriers in the land and established them there when he founded the steel and tinplate works on the Zanze." The letter referenced a report from 1768, by Jaeckel, which noted great quantities of small iron wares, "the kinds of things that peddlers sell door-to-door." "Here is the question," it continued. Did Justi have the authority to "engage and establish" those workers in the Neumark? Berlin could find no evidence of such an order, and yet, there was clear evidence that such small-scale iron workers were there, in the Neumark. Some had died out. Some had established themselves in small towns. But they were there. Now the department wanted to know "whether these *ouvriers* were in the king's pay or working on their own account, and how the sale of these small manufactures was handled at that time." It also wanted to know if the workers had "peddled them door-to-door, with passes furnished by von Justi." This prompted some serious questions. Had he been authorized to do that? And who had furnished him with the passes?[108]

The implications were clear enough. Justi had systematically defrauded the crown by creating a black market for royal iron. Thirty years earlier, Reichard, Jaeckel and others had noted the miserable account books and heaps of missing iron. No one had ever found it. Now, three decades later, there was finally an explanation: Justi had imported workers to the Neumark to sell the king's iron door-to-door for private gain. The presumption, of course, was that Justi had been paid on the side, and that he was reselling some of the Swedish and Siberian iron purchased directly by the crown. The Kammer was right. Johann von Justi was a bad cameralist.

CHAPTER FIVE

Useless Sciences, Fashionable Sciences

It was the desire to rule, the powerful craving to make people happy, that drove him. . . . Now, imagine a man of mean birth and low rank without the smallest hope of ever being able to serve in the offices of state, and yet filled with this passionate craving! But now—now this mass of disorder melted into the stream of his future vocation. "No, no! I did not want to be a ruler myself," he exclaimed when he was alone, "but to educate the servants of rulers and princes, the guarantors of the people's prosperity; that was it, and I knew it not."

—Johann Heinrich Jung-Stilling, *Autobiography*.

In 1789, a journal playfully addressed to "œconomists, cameralists, housemothers and garden-lovers" appeared in Erfurt. Its editor, an official named Barkhausen, ridiculed the myriad attempts by academics to educate farmers. He criticized scientific societies filled with "bookish scholars and œconomists ignorant about what kind of grain to plant in the winter or the summer. . . . Such gentlemen go astray with their merely speculative reasonings, which have little worth for the practical *Oekonom.*" They excelled at transforming simple knowledge into impenetrable learned babble, at using their "algebraic $X = U+A - Z$ to confuse the layman's healthy understanding." This journal would be different. It would expose the stupid errors of agricultural pedants and provide useful material for the "thinking farmer." It would communicate useful knowledge to useful people.[1]

The journal's initial article, a fictional letter from a concerned father to a well-known professor, demonstrated the dangers of learned intrusion into the realm of practice.[2] The father, a choirmaster, had sent his son Wilhelm to study law at the university. Soon, however, the young man became enamored of the œconomic and technological sciences. After learning about "technology" from Wilhelm, his father wondered how it could ever be of use.

> You could have studied farming *(Oekonomie)* at our neighbor Klausmann's, . . . and why are you troubling yourself over artisans when you have a higher calling and could someday be a town clerk? I don't see how you can call something a science (*Wissenschaft*) which consists only of skillful knacks. Science must be capable of demonstration through the Wolffian method, and demonstration is proof through indisputable grounds. The shoemaker, for example, doesn't cut his leather according to indisputable and unchangeable grounds, but today this way, and tomorrow in another way, according to the demands of fashion.[3]

The choirmaster's dismay increased as he learned about Wilhelm's entire cameralist course of study—botany, popular medicine, commercial science, mining science, forest science, and many other strange new sciences with Greek names—"so that by the end I was astonished by everything that a young person has to learn these days."

When Wilhelm completed his studies and returned home, his father sent him to the old town clerk for an examination. But the old clerk declined to examine Wilhelm, noting that the young man had clearly applied himself to "*galante Studia.*" "You are mistaken sir," the choirmaster replied, "my son has studied all possible sciences and speaks more about dung and potatoes than about writing verse." "Yes, yes," replied the clerk, "I call everything '*galante Studia*' which doesn't belong *ad praxin judicialem.*"

Unable to find work, Wilhelm began performing agricultural, botanical and technological experiments in the family's fields, in the garden, and around the house. "The *Ökonom* who has learned only through experience," Wilhelm explained to his father, "will never make great advances." Rather, those who had acquired a solid theory of agriculture, who had studied *Naturkunde* and the other auxiliary sciences, and who had gathered knowledge of foreign things would succeed. Wilhelm introduced new planting methods, purchased a wood-saving stove, suggested stall-feeding for the cattle, and sought to introduce woad and tobacco to the family's fields. Unfortunately, the innovations proved disastrous—the crops died, and the wood-saving stove smoked out the house.

After these failures, the choirmaster dispatched his young cameralist to a wise old magistrate, a good friend and relative, to learn about administrative practice. But two months later the magistrate sent him home again, explaining that Wilhelm had little notion about the theory of jurisprudence and, even worse, found the work too dry and boring to apply himself to it. "Judicial practice," explained the old man, "can never be as easy or amusing as a journal, a poem, or a philosophical-seeming treatment about the wonders of nature." His message was clear: Wilhelm had been spoiled by the œconomic and cameral sciences. "It is not the first time in the last ten years," he added, "that young people have come to me who, despite all talent and industriousness, are of no use in the business of civic life." Especially useless were those who had applied themselves to the "so-called cameral sciences," a course of study that had become fashionable and increasingly prominent since the end of the Seven Years' War. The old judge suggested that the situation at universities was out of control and that the influence of the cameral sciences was to blame. Students now had to learn not only speculative philosophy, history, and languages but also "jurisprudence, pure and applied mathematics, physics, universal natural history, chemistry, mineralogy, botany, mining science, commercial science, forest science, agriculture, technology, police science, general and particular politics, statistics, cameral sciences narrowly understood, and God knows what else."

For Wilhelm, it was too late. When his father finally sent him to a neighboring estate to become a "practical farmer," he lasted only four weeks. "Sir, take back your son!" the manager demanded. "He's ruining my entire estate. He knows everything better . . . though he can't even work a plow." In addition to ruining the crops again, Wilhelm had stabbed two cows to death while experimenting with a novel Swiss cure. Finally, convinced that his son's fashionable knowledge was as useless as all the expensive technological books and models—French plows, Dutch windmills, and other such things—acquired at the university, the choirmaster resorted to one last option. "Couldn't my son, who has heard all the cameralist lecture courses twice over, and taken careful notes, earn his bread as a private teacher and eventually secure a professorship at your university? . . . he really does know an amazing amount, just not exactly what one needs to become a town clerk or an *Amtmann*."[4]

Contemporaries would have recognized Johann Beckmann as the target of Barkhausen's satirical critique.[5] Beckmann, Göttingen's Linnaeus of agriculture and industry, had made himself famous by classifying and categorizing every occupation of the *Nahrungsstand*. People might make fun of this (some of Beckmann's own colleagues certainly did), but Barkhausen was absolutely right: the cameral sciences were above all fashionable sciences. Beckmann had succeeded by virtue of bringing attention to himself and his university.

His success inspired imitators, and the most successful of them was Friedrich Casimir Medicus, who would found an entire academy based on Beckmann's model.

Founded in 1774, and transplanted to the University of Heidelberg in 1784, the cameral academy (Kameral Hohe Schule) in Lautern[6] has been called "the most successful Cameralistic teaching institution of the eighteenth century"[7] and "perhaps the most influential of the institutions that offered the curriculum of the cameral sciences at the end of the eighteenth century."[8] With a curriculum rooted in technology and the natural sciences, Lautern's cameral academy employed permanent professors to teach subjects such as chemistry, œconomic botany, technology, and agriculture. The plan that Justi had first imagined twenty years earlier in Göttingen achieved its full realization in Lautern.

The driving force behind the cameral academy was Medicus, court physician in Mannheim. Medicus, a founding member of the Palatine Academy of Sciences, had assumed direction over the Lautern Physical-Œconomic Society in 1770. Using his connections at court, Medicus quickly managed to secure an illustrious membership for the new society. With Karl Friedrich of Baden (the champion of German physiocracy) and Bavarian Elector Max Joseph as honorary members, and with Count Karl August of Zweibrücken as its president, the Lautern society soon sparkled with the trappings of wealth, status, and fashion. Medicus even convinced the elector Palatine to take money from his academy of sciences in Mannheim and give it to the Physical-Œconomic Society.[9] During the early 1770s, the society purchased a "model farm" near the village of Siegelbach and established a manufactory in Lautern. When the cameral academy opened its doors in 1774, it was one of several enterprises owned and run by the Physical-Œconomic Society.

Though Justi had imagined an independent faculty of œconomic and cameral sciences two decades earlier, the blueprint for Lautern was Daniel Gottfried Schreber's 1763 plan for an "academy of œconomic sciences."[10] Schreber conceived of a mixed institution that would function as manufactory, farm, and academy. The academy's professors would lecture on how to direct state institutions, and they would also manage real factories and farms. Profits from the academy's crops and manufactures would support the institution. Unlike the useless professors at most universities, Schreber's academic cameralists would support themselves with something other than lecture fees. In this way, the state could train its cameralists without burdening the *Kammer*. Professors would not be permitted to wallow in empty speculation; each of them would play an active role in the life of the fiscal farm-manufactory-academy. The professor of mineralogy would teach "physical chemistry" in the laboratory and "œconomic chemistry" in the manufactory, devoting attention to dyes, metals, and ceramics. The professor of mathematics would have to teach students how to survey fields and mines. The professor of natural history would oversee the academy's gardens, fields, and nurseries. In short, Schreber envisioned a state in miniature where aspiring cameralists could practice their skills.

Between 1770 and 1777, Medicus worked to realize Schreber's model. He assumed direction over the Lautern Physical-Œconomic Society in 1770, oversaw the initial establishment of a linen manufactory in 1771, and supervised the acquisition of the Siegelbach estate in 1772. In 1774, with lectures scheduled to begin at the new academy, Medicus engaged Georg Succow, a young physician from Jena, on the strength of a recommendation from Schreber. Succow assumed responsibility for an array of "basic sciences" (*Grundwissenschaften*)—chemistry, mathematics, natural history, and natural philosophy. In the following year, the academy hired Ludwig Schmid to teach state administration and fiscal sciences proper. Finally, Johann Heinrich Jung-Stilling arrived in Lautern in 1778 as professor of technology, trade, and agriculture.[11] By 1778, Medicus had assembled all of the elements envisioned by Schreber fifteen years earlier. He had purchased a farm, acquired a manufactory, and hired professors to teach the entire range of cameral sciences.

Like Schreber, Medicus believed that professors of the cameral sciences ought to be skilled, practicing cameralists. Above all, this meant that they were expected to have experience in managing productive enterprises like manufactories, farms, and mines. Jung-Stilling, for example, clearly felt that his practical experience had prepared him to lecture on the cameral sciences.

> "Who can teach it better than I?" thought he to himself. He had lived long in the woods, amongst foresters, charcoal burners, woodcutters, etc., and was therefore perfectly acquainted with the practical part of these things. Surrounded from his youth up by miners of every description—with iron, copper, and silver smelters, with bar-iron, steel, and spelter-founders, with wire-drawers—he had become thoroughly acquainted with these important manufactures and had also managed estates and foundries for a total of seven years at Mr. Spanier's. At the same time, he perfectly understood commerce in all its branches, and was experienced in all of them. And in order that he might not be deficient in the fundamental and auxiliary sciences, Providence had very wisely directed him to the study of medicine, in which physics, chemistry, natural

> history, etc. are indispensable. In reality, he had labored through these sciences, and especially mathematics, with greater fondness than all the rest, so that even in Strassburg he had read a lecture upon chemistry.[12]

Jung-Stilling's description included not a word about his aptitude for political economy, politics, or state finance. Rather, it was his detailed knowledge about the sources of wealth, about *Nahrungsquellen,* that qualified him to teach at Lautern. For ambitious professors like Jung-Stilling, the cameral sciences became the vehicle through which the prestige and power of governance suffused the natural sciences. They offered the only path by which a man like himself, of "mean birth and low rank," could satisfy the craving to rule.

DIRTY, BACKWARD LAUTERN

When Medicus became director of the Lautern Physical-Œconomic Society in 1770, Lautern (Kaiserslautern) was home to one of the poorest districts in the Palatinate.[13] For the elector Palatine and his ministers, the region was an embarrassment. Contemporary descriptions of the Lautern district—nonexistent manufactures, scraggly forests, stagnant streams, shabby farms, dilapidated houses—bear many similarities to the accounts of 1720s Göttingen. When Jung-Stilling saw Lautern for the first time in 1778, for example, he was struck by the town's "old irregular houses, low rooms, with ceilings supported by crossed beams, little windows with round or hexagonal panes of glass, doors which could not be shut, stoves of dreadful dimensions, . . . then a view into nothing but gloomy pine forests; nowhere a rushing stream, but snaking, creeping, marshy water."[14] For an improvement-minded cameralist like Jung-Stilling, Lautern's houses, stoves, streams, farms, and forests showed the physical signs of stupidity, laziness, and neglect.

Jung-Stilling joined a group of improvers who had, with the support of the electoral government, been struggling to initiate change since the 1760s. Led by the apothecary Johann Riem, the group had founded a Lautern Physical-Œconomic and Bee Society in 1769.[15] The bee society, part of a larger movement that spawned many œconomic and patriotic societies in the empire after 1760, dedicated itself to the improvement of regional agriculture.[16] Inspired by the physiocrats, Riem believed that regional prosperity depended on agriculture. But it was Medicus who would become most important for the future direction of the society.

Medicus arrived in Lautern via Mannheim, where he had managed to court the favor of Elector Karl Theodor. Between 1765 and 1767, he planned and installed the ground floor of Mannheim's botanical garden.[17] He also dedicated himself to the improvement of regional agriculture and became known through his essays on "practical botany" in the *Pfälzischer Landkalender.*[18] The Palatine Academy of Sciences had developed the *Landkalender* as a way to educate the land's "country people" about good agriculture. "One finally realized," Medicus recounted later, "that one had to make the country man (*Landmann*) aware of the main principles of agriculture upon which the prosperity of the state rests."[19]

Medicus used his essays in the *Landkalender* to sing the virtues of stall feeding, clover, cow manure, and dung of all kinds. "You dear country people!" began one essay, "you know all too well, that dung is the soul of agriculture." The tone of the essays was straightforward, folksy, condescending, and sometimes peevish. Medicus often complained that the region's farmers would prosper if only they followed good principles. Agriculture was flourishing in Württemberg and Swabia, he claimed, because farmers there understood the importance of dung. Medicus had seen it for himself. "The farther I traveled into Swabia," he wrote, "the more often I found dung spread on the pastures, and it was amazing how much dung one spread about every winter on the pastures in the region around Aalen, Schwäbisch-Gmünd, Schorndorf and surrounding areas."[20]

It is tempting to see Medicus's "practical botany" and his popular writings in the *Landkalender* as the by-products of French physiocracy.[21] But Medicus was not a physiocrat. Though he believed in the importance of agriculture, he treated it as one element in a larger productive system. In fact, Medicus justified his intensive brand of agriculture in the same way other cameralists justified woolen manufactories or silver mines. The Swabians, he liked to explain, had bred large herds of cattle because of their excellent use of dung and stall feeding. As a result, they had attracted "an extraordinary amount of foreign money" into Swabia through the cattle trade. If Palatine farmers followed Swabia's lead, the region's cattle might become a source of surplus rather than loss.[22] For Medicus, introducing new farming techniques was just like encouraging new manufactures. In both cases, state officials aimed to increase the prosperity of a larger productive system and to enrich the Kammer by establishing individual enterprises.[23]

THE MODEL FARM

In 1772, the Lautern Physical-Œconomic Society bought an estate in the village of Siegelbach for two thousand florins.[24] The society hoped this "model farm [*Mustergut*]" would become a center of agricultural reform and experimentation. It included about sixty-six hectares of land, along with houses and farm buildings.[25] Because the society had spent all its money on the estate, it had to lease the Siegelbach property to a tenant (*Pächter*) until it could raise some capital. When the cameral academy opened its doors in 1774, the society finally took over administration of the model farm, but there was still not enough money to buy cattle, seed, and tools. To raise the necessary capital, the Lautern society offered seventy-two shares in the Siegelbach farm. It managed to sell twenty four of them, raising two thousand florins for the enterprise. The remaining forty-eight shares reverted to the society, which now hired an administrator to run the estate and appointed a commission to oversee the operation. The share offering was similar to the distribution of *Kuxen* in the mines of the Harz and Erz mountains. In Siegelbach, where there was no silver, investors hoped that flax, madder, clover, and cattle would yield rich returns.

The Lautern Physical-Œconomic Society had been holding regular meetings and publishing essays dedicated to the improvement of Lautern's agriculture for years. Since 1769, it had also owned a plot of land in Lautern for its botanical garden.[26] With the acquisition of Siegelbach, the society could demonstrate the effectiveness of its principles by reviving a barren estate. Its members regarded Siegelbach, quite literally, as a model for Lautern and the Palatinate. It would, they hoped, represent in miniature what the region's fields and farms could become, if only they were properly managed and directed. Hopes were high in 1774, as the society took over administration of the model farm.

From the beginning, things did not go well. The Siegelbach property was divided into many different plots, making it difficult to farm. Moreover, the land was relatively infertile. The society tried to put Medicus's notions into practice by fertilizing with lime and gypsum, planting clover, and working to collect dung from the small Siegelbach cattle herd. The cows were kept from grazing and fed in their stalls, the pigs were bred intensively, and there were experiments with horse breeding. The society had individual stalls built for all forty-eight head of cattle, and it purchased another fourteen acres of land to fill out the property. The distance between Lautern and Siegelbach, however, began to create problems, since the society's members had little control over the local administrator, Herr Saner. Over time, capital for the enterprise dried up. There was insufficient dung to increase the feed, and insufficient feed to increase the dung. By the end of the 1770s, the society's model farm was in danger of failing altogether.[27]

When Jung-Stilling arrived in Lautern in 1779, he was sent to Siegelbach. None of the other administrators had been able to make it work, but Medicus hoped that his new professor of agriculture and technology, armed with systematic knowledge, would make the place turn a profit. As Jung-Stilling recalled years later, the connection between his teaching post and the management of Siegelbach had been explicit. As the academy's "teacher of agriculture," he accepted this "subsidiary office" willingly, because he thought he could make it succeed. Medicus expected Jung-Stilling to make the model farm thrive. It would provide revenue for the society, instruction for the academy's students, and an example for the whole region. At Siegelbach, one would "make all kinds of agricultural experiments and guide the farmers with good examples."[28] Jung-Stilling took over the model farm immediately upon his arrival in Lautern.

When he traveled to Siegelbach, Jung-Stilling found an impressive cowshed built with large blocks of stone and constructed "according to the new fashion."[29] Unfortunately, he also discovered that it housed "twenty scrawny skeletons of Swiss cows that all together yielded three pints of milk each day." In addition to the scrawny cows, the farm had two workhorses, two foals, and a number of pigs. Though it was only November, all the hay and straw was long gone, and the household had no stores of butter, milk, or feed. Jung-Stilling was shocked. He turned to the society for help, but its members did not want to waste more money on Siegelbach. They told him to manage with what he had.

Jung-Stilling now committed himself to making Siegelbach a success. He felt that his own honor, and the honor of the cameral academy, depended on Siegelbach's success. He decided to sell half the cattle, hoping that he could use the proceeds to buy straw and feed for the other half. But at the official cattle auction the professor realized, to his horror, that the audience was filled with Siegelbach's creditors. The auction went badly, and Jung-Stilling, who was already in serious debt, began investing his own money in the place. He borrowed money to make the farm succeed. He decided to plant fields of clover and hay as feed for the animals, convinced that his investments would yield surpluses by the following summer. Persuaded by the principles that he and Medicus had been promoting in the publications of the Physical-Œconomic Society, Jung-Stilling believed that a firm commitment to stall feeding and dung application—"dung is the soul of agriculture"—would turn Siegelbach around.[30] Things did not work out that way. As Jung-Stilling recalled later, "everything went wrong." In the end, he gave up the administration of Siegelbach and retreated in failure, believing that unfaithful servants, thieving neighbors, malicious subaltern officials, heavy debts, and a general lack of support had ruined the enterprise.[31]

In 1781, after Jung-Stilling's disastrous tenure as administrator, the society leased the property to a tenant, with the proviso that he adhere to the "principles and purposes" of the Physical-Œconomic Society.[32] The tenant, who apparently had different principles and purposes, pocketed the money that the society had advanced to him, pawned everything that he could get his hands on, and fled for Hungary. After that, the society decided to cut its losses by selling the Siegelbach estate. The members drew an important lesson from the model farm fiasco. "The greatest secret of farming is the watchful eye and the uninterrupted oversight of the owner, and where this is impossible, the result will be nothing but losses instead of profits."[33]

LAUTERN'S MANUFACTORY-WORKHOUSE

Years earlier, when Karl Theodor first became elector of the Palatinate in 1743, he was only eighteen. His predecessor, Karl Philipp, had moved the court from Heidelberg to Mannheim and had begun work on a new palace and grounds to accommodate the court.[34] Then, in 1743, the young elector turned his attention to the construction of court and town. Karl Theodor demonstrated an insatiable appetite for entertainment, mistresses, and other elegant symbols of power. He built grand new buildings. He supported the theater, funded the Palatine Academy of Sciences, established pleasure gardens, bankrolled scientific societies, erected libraries, and invested in paintings and sculpture. Contemporaries soon recognized Mannheim as a center of art, music, architecture, and the sciences. An observatory was built for the renowned Jesuit astronomer Christian Mayer. Voltaire, that consummate enlightened moocher, even came to stay for awhile.[35]

Since 1720, when Karl Philipp first moved the court from Heidelberg, Mannheim had become a center of projects and enterprises.[36] The palace, whose construction almost single-handedly bankrupted the Palatinate, was long a focus of activity. One needed builders and artisans, butchers and bakers, natural philosophers and musicians. Karl Philipp, in turn, embraced projects that he thought might benefit the region, but not all were successful. The elector's decision to grant a tobacco monopoly and a loan of one hundred thousand florins to the Spaniard Don Bartolomeo Pancorbe de Ayala y Guerra, for example, was a complete disaster. Nor did the establishment of a government trading house fare much better.

When Karl Theodor came to power, Mannheim and the surrounding regions boomed as never before. The elector's appetites created demand for luxury goods of all kinds. But Karl Theodor's lavish court caused headaches for the Kammer because it was enormously expensive. Before long, the constant demand for luxury goods in Mannheim led the government to support local manufacturing ventures. It would be far better to supply the court in this way, reasoned the Palatinate's fiscal officials, than to waste precious silver on foreign wares. Entrepreneurs, looking to profit from the new opportunities, founded many enterprises, and they almost always received support from the authorities. Karl Theodor's "commerce commission," established in 1768, supported new manufactures and encouraged trade. The Mannheim authorities also poured thousands of florins into Frankenthal, which soon became a center of state-sponsored manufacture. Frankenthal boomed between 1760 and 1778, and, with the help of the Kammer in Mannheim, manufactured goods began sprouting up there—porcelain, woolens, silks, playing cards, tapestries, starch, candles, and so on.[37]

Initially, Lautern did not benefit much from Karl Theodor's many initiatives. It remained a backward region, physically isolated from the thriving centers of Mannheim and Frankenthal. The region's troubles were exacerbated by heavy tax burdens, since Karl Theodor and his ministers supported the lavish style of the court largely on the backs of peasants and farmers. With the establishment of the Lautern Physical-Œconomic Society in 1769 and the rise of Medicus to the directorship in 1770, however, Lautern forged connections with Mannheim. The society, and especially Medicus, hoped to integrate Lautern into the larger system of commerce and manufacture that emanated from Mannheim. The elector, always on the lookout for profitable ventures to support, embraced the efforts of the fledgling society.

From its earliest days, the Physical-Œconomic Society had considered ways to keep the region's peasants busy. The members were annoyed that so many of the country people sat idle for much of the year. In 1770, Dr. Stefan Gugemus, a member of the society, proposed that the government establish workhouses so that Lautern's idle residents could knit, spin, and weave during the winter months.[38] Other members of the society embraced Gugemus's plan. Lautern's town treasurer, Ludwig Fliesen, supported the venture, and Medicus eventually convinced the elector to back it. On 18 April 1771, Karl Theodor promulgated statutes and privileges for a new manufactory and workhouse in Lautern.[39]

The electoral patent gave the society a monopoly over the production of linen, "half-linen" and "siamoise" in Lautern.[40] Mannheim encouraged the enterprise in terms that Justi himself might have written. The success of "new factories and manufactories," explained the patent, could bring prosperity to the lands of the Palatinate by making its residents more productive and useful.[41] When Medicus and the Physical-Œconomic Society proposed to establish a new linen manufactory in the town, the authorities, who were concerned about the "greater improvement of our *Stadt* Lautern," embraced the plan. Mannheim promptly granted all the special "privileges, freedoms, and immunities" that the society had requested.

The society's manufactory was really a state-sponsored workhouse. Mannheim provided a manufacturing house, gave the workers amnesty from military service, offered tax freedoms, and granted exemption from guild regulations. Moreover, the electoral government ensured that the Lautern manufactory would have a ready supply of labor power, for it promised to round up loafers and send them to work there. "The idle and derelict," declared the patent, "shall be collected by the *Oberamt* and sent to work at the factory for a suitable wage." The Lautern Physical-Œconomic Society, originally founded as a bee society and dedicated to the improvement of regional agriculture, had now become a "society for manufactures" as well.[42] Its members, in turn, had become the directors of a workhouse.

Though the head of the manufactory was a local merchant named Philipp Heinrich Karcher, the "manufactures commission" also included fiscal officials, foresters, jurists, and even an army colonel. By 1774, when the cameral academy began offering courses, the linen manufactory was already employing nine weavers, four cotton spinners, eight spoolers, and forty-three women as flax spinners.[43] Much like Grätzel in Göttingen, the manufactory directors soon recruited cheap labor from throughout the countryside to spin and weave in their homes for the manufactory. Members of the Physical-Œconomic Society focused on improving and finishing the product. In 1772, they acquired a dyeing works, and the co-director of the cameral academy, Christian von Hautzenberg, took over the bleaching works in 1775. The society's linens and cottons sold well, and the enterprise eventually became a success, employing 1,810 people by 1784.[44]

Things had not started so well, and Medicus initially feared that the Lautern manufactory might go the way of the model farm in Siegelbach. In 1776, he had asked Jung-Stilling (who was not yet a Lautern professor) for details about the manufactures in Strasbourg.[45] Textiles were thriving there, and Medicus wanted to know why. He asked Jung-Stilling to report about production techniques, the organization of manufactures, and how the manufactory in Lautern might be improved. Jung-Stilling, however, felt that it was too dangerous, since he might be taken for a Palatine spy. He wrote to Medicus, explained his fears, and offered to write treatises for the society instead. Jung-Stilling prepared one treatise after another, and Medicus read them before the members of the society.[46] These treatises on manufacturing techniques testified to his qualifications for the professorship in Lautern. Like any good informant for the Kammer, he had demonstrated his ability to judge local manufactures based on systematic knowledge and wide experience. In other words, he had embodied the virtues of a good cameralist.

Ironically, the success of the Lautern's linen and calico manufactory had very little to do with the scientific efforts of the Physical-Œconomic Society or the learned professors at the cameral academy. Rather, it was the merchant, Karcher who transformed the operation into a success. Like Grätzel in Göttingen, Karcher had a knack for dyes, and he was good at making the kind of calicoes that appealed to customers. Moreover, he managed to attract skilled workers to Lautern and to train others, no small feat in a place with such a bad reputation. The members of the cameral academy and the society, who had put up much of the capital for the enterprise, had the luck to invest in a successful venture.[47]

THE CAMERAL ACADEMY

With the help of Palatine fiscal officials, Medicus had begun to build a system of institutions in Lautern. The society discussed and debated projects at its meetings, conducting experiments with new plants and crops in the botanical garden and trying to cultivate them, without any success, at the model farm in Siegelbach.[48] The society's members considered how to cultivate flax and produce dyestuffs for the manufactory. They looked for the best methods of keeping Lautern's country people busy during the winter months. In fact, with its regular meetings, formal presentations, debates over projects, and collegial voting procedures, the society mimicked a Kammer collegium. Its members were playing at fiscal governance.

The cameral academy began as a venture, much like the farm and the manufactory. Medicus decided that the crown needed qualified cameralists to manage the Palatinate's farms, factories, and mines. Looking to other successful academic enterprises in Göttingen and Freiberg, he also saw the potential to make some money from the cameral sciences. In April of 1774, the society resolved to open a cameral academy in Lautern, and it petitioned Mannheim for permission. Karl Theodor gave the idea his blessing, and the Kameral Hohe Schule opened for business on 14 October 1774.[49]

Medicus knew that the success of his new academy depended on sales, as it would for any infant enterprise. One had to stave off competition and secure a market for the product. The Lautern siamoise manufactory, for example, depended largely on the electoral patent, which granted it a regional monopoly over the production of certain cottons and linens. For the cameral academy, competition came mainly from the University of Heidelberg, traditional alma mater of the Palatinate's educated fiscal officials. It was, therefore, crucial to stifle competition from Heidelberg (and other universities), while promoting the scientific wares of the new academy.

Medicus began an advertising campaign soon after the academy opened for business. He directed his new professors, Succow and Schmid, to send articles about the Lautern academy to important journals and periodicals throughout the Empire.[50] He contacted Isaak Iselin in Switzerland, asking him to promote the cameral academy in his new journal, the *Ephemeriden*.[51] He sought out new markets for the Physical-Œconomic Society's periodical, the *Bemerkungen*.[52] In general, Medicus did everything that a diligent manufactory director would do to find new markets.

By 1778, his constant lobbying for the college yielded dividends. In December of that year, the Hofkammer in Mannheim ordered that no one would be appointed to upper-level fiscal positions (*Kameral-Obere Stellen*) or state offices unless he had studied at the cameral academy in Lautern.[53] The academy had been established to train "cameral and administrative officials" for the Palatinate. The intention, as the regulation explained, was to train cameralists who could make the land prosper. The Kammer, in turn, would reap the "benefits" that flowed from a wealthy people. Unfortunately, these expectations had been disappointed since "said academy has been attended by foreigners, but not by native *Pfälzer*." As a result, the Hofkammer now ordered that all candidates who hoped to work for the Kammer would need evidence of their "industriousness" from the professors in Lautern.[54]

Medicus had used his considerable personal influence with Elector Karl Theodor and Minister von Oberndorff to secure a monopoly over fiscal-administrative education, and the 1778 ordinance gave him immediate leverage over desirable appointments to state office. Not surprisingly, there was significant resentment and resistance to the new policy. For some fiscal officials, like the tax collector Marc Anton Faßele in Mannheim, the new ordinance came as a shock. Faßele lobbied for an exemption, explaining that he had been a good cameralist for ten years, and that he did not need professors in Lautern to teach him his business.[55] Fiscal officials in the Palatinate, many of whom resented Medicus's intrusion into their affairs, sometimes neglected the privilege or failed to enforce it. In 1779, for example, the Hofkammer issued a reprimand, since it had come to the attention of the authorities that many fiscal officials were ignoring the ordinance.[56] Medicus, however, doggedly defended his monopoly. Between 1778 and 1784, as fiscal officials petitioned the Hofkammer for exemptions from the ordinance, Medicus consistently argued for its rigorous enforcement.[57]

Mannheim granted Medicus his monopoly for many of the same reasons that Hannover had given Johann Heinrich Grätzel his Camelott patent in Göttingen. Like every young venture, Lautern's academy was a "delicate plant" in need of care and encouragement. The 1778 privilege, like a ban on cotton goods or foreign linens, forced buyers (that is, candidates for state office) to purchase their professional education in the region, since training at foreign universities, or even at Heidelberg, was no longer a path to positions in the Kammer. Moreover, the privilege was intended to bring wealthy students to Lautern, where they would spend their money. The authorities in Mannheim hoped that Lautern's cameral academy, like the University of Göttingen before it, would help revitalize a backward town.

Elector Karl Theodor visited Lautern's cameral academy for the first time on 30 September 1777. Accompanied by Minister von Vieregg and General von Belderbusch, the elector inspected the collections, holdings, and properties of the new academy. After touring the library, the natural history collection (*Naturalienkabinett*), the model collection, and the physics instruments, the elector and his companions moved on to the manufactory, which was in the same building. They inspected "two stores of raw and spun wares." From there they toured the spinning rooms, the printing works, the coloring area, the weaving rooms, the chemical laboratory, the dyeing works, and finally the vault containing finished wares.[58]

Medicus ran his new academy like a manufactory, and it really *was* a manufactory. The various subsidiary enterprises of the Lautern Physical-Œconomic Society—model farm, cotton and linen manufactory, cameral college—were part of a single system. Together, they formed an idealized cameralist state in miniature. The curriculum at the academy aimed to show aspiring cameralists how to make such enterprises succeed. There was only one problem: the academy succeeded for all the wrong reasons. At the model farm in Siegelbach, all the fashionable new cameral sciences—the "galante Studia" that Barkhausen spoofed in his 1789 satire—proved completely useless. At the manufactory, it was a practiced merchant, and not a professor, who made things work. Most important of all, the cameral academy succeeded by selling its sciences to the public and by establishing an academic monopoly in Lautern. That was the most important lesson that Medicus learned from Münchhausen's Göttingen and Heynitz's Freiberg.

MEDICUS'S PROPOSAL FOR INGOLSTADT

While working in the Generallandesarchiv in Karlsruhe, I stumbled on an extensive proposal for a new "faculty of state administration (*Staatswirthschaft*)" at the Bavarian University of Ingolstadt (see appendix 3).[59] Initially, I assumed that Franz Xaver Moshammer, professor of the cameral sciences in Ingolstadt, was its author. Moshammer, a student of Göttingen's Johann Beckmann, taught the cameral sciences in Bavaria for many decades, and his published plan for Ingolstadt is well known.[60] But in 1784, when this secret plan was drafted, Moshammer was too young and too scared to write anything of the sort. He did not even dare to publish his own textbook. As he explained to his friend and mentor Beckmann, outspoken professors risked "great unpleasantness and danger in my fatherland, where one is not at all used to publicity (*Publicität*)." Moshammer preferred to keep his head down.[61]

In fact, Friedrich Casimir Medicus, director of the cameral academy in Lautern, drew up the plan.[62] He drafted it in February 1784 while in residence at Karl Theodor's court in Munich.[63] The elector Palatine had moved his court to Munich in 1779, when he became ruler of Bavaria after the death of Elector Max Joseph III in 1777.[64] Karl Theodor felt like an outsider in Bavaria (which he was), and he felt surrounded by suspicion and intrigue there (which he was). All of this led to a strange mix of policies in Bavaria. On the one hand, he tried to introduce many of the same administrative reforms that he had used in the Palatinate. On the other hand, the considerable influence of his confessor, the ex-Jesuit Father Ignaz Frank, led to a stubborn fight against many of the things that Karl Theodor, the theater-loving, opera-going, loose-living "first cavalier of the Holy Roman Empire," represented.[65]

After 1778, when the elector and his court left for Munich, officials in Mannheim assumed more direct control over domestic affairs in the Palatinate. Friedrich von Oberndorff became the prime minister and head of the Kammer in Mannheim during that year.[66] The new position allowed Oberndorff to become the Münchhausen of Lautern. His name is everywhere in the rescripts, orders, memoranda, and other paperwork that flowed between Mannheim, Lautern, and Heidelberg during these years. Moreover, professors at the academy acknowledged Oberndorff as their patron and protector.[67] Karl Theodor continued to rely on many of his trusted advisers from the Palatinate as he took over the government in Munich.[68] Medicus benefited from the elector's trust. By February 1784 he had successfully courted the favor of both the elector and the Palatinate's prime minister. This is probably why he found himself in Munich during the winter of 1784, drafting a plan for a new faculty of state administration.[69]

As Medicus drafted his plan for Ingolstadt in 1784, rumors were afoot about the imminent move of the cameral academy to the University of Heidelberg. (The Kameral Hohe Schule, with all its professors and collections, would indeed move to Heidelberg later that year.) Medicus's plan for Ingolstadt, therefore, is one of the most comprehensive overviews we have of the "Lautern system" from the man who established it.[70] "Ten years of reflection and the direction of the Lautern College," he wrote, "have acquainted me with this subject, and with everything that is necessary for it; based on my experiences, I would like to outline my thoughts about the creation of this new faculty."[71]

Medicus outlined a full course of study for future cameralists in Ingolstadt. He divided the curriculum into three major parts: foundational knowledge (*Grundlehre*), knowledge of resources (*Quellenlehre*), and actual state administration (*eigentliche Staatswirthschaft*). For Medicus, the success of the entire system rested on the "foundational sciences" (*Grund Wissenschaften*). These included natural history, pure and applied mathematics, physics, chemistry, and mineralogy. Others, like Beckmann and Justi, had called these the "auxiliary sciences." But Medicus, who considered the natural sciences too important to be relegated to the margins, opted for a name that reflected their importance. It was crucial, he believed, to have one or two professors dedicated entirely to the natural sciences. Succow had filled this position in Lautern for ten years. The *Grund Wissenschaften,* Medicus claimed, were the "true basis upon which all the knowledge of a future state administrator rests, and without which he can never make a sure step forward. One must therefore guarantee that no pupil who has failed eagerly to study these sciences is allowed to proceed to the resource sciences (*Quellen Wissenschaften*)."[72]

These so-called resource sciences aimed to classify and examine all sources of wealth within the state. Medicus divided them into agriculture, forestry, mining, and trade. Though he used a different term here, Medicus was referring to the same group of descriptive and classificatory subjects, such as *Technologie* and *Handlungswissenschaft,* that Johann Beckmann had made so fashionable in Göttingen. Jung-Stilling taught these subjects in Lautern. In fact, Medicus warned that the resource sciences should have no more than one teacher. Otherwise, one risked the fate of the "French œconomists" (the physiocrats), who had improperly raised one part, agriculture, above the rest. The true *Staatswirth,* by contrast, preferred no one part to the others, for he had to direct the whole—mines, manufactures, forests, and farms. Medicus recommended one of Lautern's own students, Engelbert Martin Semer, as the most qualified candidate for teaching the resource sciences at Ingolstadt.[73]

After thorough study of the natural sciences and the "doctrine of resources," candidates could move on to the final stage of training. They were now prepared for initiation into the mysteries of state administration proper, which included three parts: police science, financial science, and *Staatswirthschaft.* Police science, Medicus explained, would teach students how to direct all the different occupations of civic life. In other words, police science would use the knowledge gathered by the resource sciences to direct the *Nahrungsstand* in the best possible way. Financial science, meanwhile, taught how to collect money from the *Nahrungsstand* to support the court, the military, and the civil service. Financial science, like the science of bee keeping, concerned itself primarily with extraction, dictating how to collect revenue without harming the source of that revenue. Candidates who survived the entire course of instruction—foundational sciences, resource sciences, police science, and financial science—would be allowed to hear lectures on "true" state administration. This final course, which drew on all the others, would instruct young state administrators in the art of governing. Medicus warned against jumping straight to the end without studying the basic sciences and the resource sciences. If the system was to work, it would be necessary to restrict entry to the later courses. If Medicus had his way, students in Bavaria and the Palatinate would hear the entire course of lectures, or they would hear none at all; every high-level state official would be trained in chemistry, mineralogy, physics, mathematics, mining, agriculture, forestry, and manufactures. Medicus also sketched out a curriculum for Ingolstadt, which he included with the plan. It established a steady progression from basic sciences to resource sciences to state sciences.[74]

The "Lautern system," as Medicus described it here, was a variation on the Göttingen model. As we have seen, Münchhausen worked with Justi, Beckmann, and others to build a shadow cameralist faculty between 1750 and 1770. Members of Göttingen's philosophical and medical faculties, like Gmelin and Achenwall, had helped to create a system of auxiliary sciences for aspiring cameralists. In Lautern, Medicus refined Göttingen's model by separating the cameralist faculty completely from the university and its older faculties. It was an innovation with significant implications for the natural sciences. At universities, sciences like chemistry and botany generally fell under the jurisdiction of the medical and philosophical faculties. Even in Göttingen, where Münchhausen did much to break up academic monopolies, the traditional faculties could and did insist on their privileges to teach certain sciences. After all, academic competition meant lost lecture fees. That is why Beckmann's efforts to teach œconomic botany met with stubborn resistance. By establishing the cameral academy in Lautern as a kind of free-standing professional faculty, Medicus avoided these problems. Like Heynitz in Freiberg, he could structure the curriculum as he pleased, without fear of reprisals from cranky medical professors.[75]

But the move to Heidelberg did take place in 1784, and it seems reasonable to wonder why. Historians have mostly relied on published advertisements, like the 1784 *Nachricht an das Publikum,* to explain the move from Lautern to Heidelberg. Keith Tribe, for example, has plausibly suggested that Medicus arranged and supported the move to Heidelberg because he "had always aimed at university status for his endeavors."[76] The unpublished "Plan for Ingolstadt," however, indicates that, as late as February of 1784, Lautern's director had no intention of moving to Heidelberg. As he wrote about Ingolstadt that winter, Medicus was already worried about the prospect of a move to Heidelberg, and he used the report to lobby against it. "When the academy in Lautern was founded," he wrote, "I believed, as I still do, that it had to be established in some isolated place." It was still much too early, he argued, to move the academy to Heidelberg because "the science has not yet developed sufficient roots, and I worry that it could be destroyed through incorporation by those who secretly oppose and envy it." By removing the school from the "native freedom" of Lautern, the electoral authorities risked destroying it.[77]

Medicus believed that the very success and survival of the state sciences, or *Staatswissenschaften,* depended on maintaining the autonomy and independence of the cameral academy.[78] One would have to avoid moving it at all costs. The academy's special curriculum had taken shape in Lautern, he explained, and it would also find its permanence there. Medicus claimed that the new faculty in Ingolstadt would "encounter almost no trouble" if the Kameral Hohe Schule stayed in Lautern. If, however, the authorities decided to move the academy to Heidelberg, Medicus predicted that the further introduction of the cameral and state sciences in Karl Theodor's lands would encounter great difficulties. The reason was simple. University professors, driven by envy and greed, would crush the new cameral institutes and faculties through their many intrigues and "cabals." The only hope for the cameral sciences, Medicus believed, was the constant threat of direct intervention by the state.

Though university professors could not be cured of their envy, they might be frightened into submission by the elector and his ministers.

HEIDELBERG

When the cameral academy moved to Heidelberg in 1784, university officials, annoyed at the special privileges of the new faculty, tried in every way to restrict its autonomy. But the new "school of state administration" enjoyed the favor and protection of the elector and his officials. Officials in Mannheim had planted an informant named Wrede in Heidelberg, and they used him to gather information on the success of the new school from Lautern.[79] Wrede praised the new professors. They had achieved widespread approval and drawn many students through the "attractive presentation of their lectures." Wrede assured the authorities that the new school would help to revive the old university and make it flourish once again. Wrede's report indicates that the decision to move the cameral academy formed part of a larger plan to reinvigorate the University of Heidelberg. Moreover, the language he used—the university would once again "come into full bloom"—echoed the frequent ministerial reports dedicated to the encouragement and establishment of manufacturing ventures. Universities "bloomed" when students purchased what the professors were selling, and, according to Wrede, Heidelberg's students liked what Medicus, Jung-Stilling, Succow, and Schmid had to offer.

Wrede was not as positive, however, about the "œconomic society."[80] Its only function was to host public lectures once a month on topics chosen by Medicus and the professors. Wrede complained that it was simply not "practical." Successful œconomic societies needed members who were experienced estate owners and farmers. Moreover, the society should be organized like those in England, so that members would present actual experiments, discuss new agricultural techniques, and present their results to the public.

As Wrede's report suggested, however, the school of state administration did not go over well with Heidelberg's established faculties. Medicus had made every effort to transform Lautern into a fashionable institution, just as Münchhausen had shaped Göttingen. In Lautern, as in Göttingen, the key was to draw wealthy and elegant students. Wealthy students from powerful families wanted privileged access to state office. This was the most valuable product that Lautern's academic factory offered. Because of his connections with Elector Karl Theodor and Minister von Oberndorff, Medicus managed to link the cameral academy with access to state office. Moreover, the special treatment from the elector, prominent nobility, and important officials had transformed backward Lautern into a fashionable address. In the 1780s, students had traveled from all over to study there. The Landgrave of Hessen-Darmstadt had even made entry to state office in his lands contingent upon study in Lautern. Now, with the move to Heidelberg, state officials hoped that the "Lautern system" could reinvigorate their backward university.

EPILOGUE

When the University of Heidelberg celebrated its four hundredth anniversary in 1786, it was the new school of state administration that captured much of the attention. As Jung-Stilling recalled later, the anniversary drew "a great multitude of people from far and near."[81] Students and visitors sat in the dank, frigid, great hall of the university, listening to "endless Latin speeches" by members of the theology, law, medicine, and philosophy faculties. Finally, the time came for Jung-Stilling to speak on behalf of the school.

> When it came to Stilling's turn, the whole audience was conducted into the hall of the School of State Administration,[82] which was a beautiful one, and as it was evening, was warm and illuminated. Now he stepped up and gave a talk in German, with his usual animation. The result was unexpected; tears began to flow—people were happy, a whisper ran through the assembly—and at length they began to clap and shout "Bravo!" so that he had to stop until the noise was over. This was repeated several times; and when he descended from the podium, the representative of the elector, the Minister von Oberndorff, thanked him very expressively; after which the lords of the Palatinate, in their stars and orders, approached to embrace and kiss him, which was also done by the principal deputies of the imperial cities and universities.[83]

Jung-Stilling, long tortured by debts and the constant threats of nagging creditors, had finally made good.[84] Looking back upon his life, he saw the 1786 speech as a turning point, for it caused the "higher ranks of the Palatinate" to notice him, thereby creating the possibility for future academic and financial success.

Jung-Stilling's speech was an extended advertisement for the new School of State Administration. But it was probably the style and delivery of the speech, as much as its content, that garnered approval from the listeners. Perhaps the audience also shared Jung-Stilling's excitement about the novelty of the thing. He claimed, for example, that "for the first time the Muse of *Staatswirthschaft* sits like a young benevolent goddess among her four sisters!!!"[85] Technically, of course, this was not correct, since the School of State Administration did not have the juridical standing of the four traditional faculties. Nevertheless, there was something true in Jung-Stilling's claim, because Medicus had successfully managed to maintain a measure of autonomy for the new institution. In any case, Jung-Stilling used the school's institutional immaturity to great rhetorical advantage. He dedicated his speech to the "young muse" of state administration, arguing that the time had come for her to take a place next to her "ice-gray sisters." Given the troubles that Heidelberg had experienced in recent years—falling enrollments, the disenchantment of the authorities, bad press—Jung-Stilling's remarks about the "ice-gray" faculties must have summoned up images of decline and decrepitude.

Nor was Jung-Stilling especially politic about the accomplishments of the traditional faculties. Had the theology faculty made the people more pious? Had the law faculty succeeded in eliminating disputes? Had the medical faculty banished disease? Had the philosophy faculty made people wiser? Why then, asked Jung-Stilling, should one expect so much more from professors of state administration? And yet Jung-Stilling suggested that faculties of state administration, unlike the older faculties, would indeed accomplish their aims by bringing wealth and prosperity to the state. No longer were battle scars and well-conducted learned disputes the only sources of honor. No longer could the prince make do with warriors and jurists as his only servants. Now, honor came to those who sacrificed their labor and resources—"the countryman his sheds and barns, the artisan and manufacturer his workshop, and the merchant his capital"—for the good of prince, people and Kammer.[86]

How they must have hated him, all those professors of theology, law, medicine, and philosophy. Their own courses were sparsely attended, and the upstart new school of *Staatswirthschaft* had increased their misery by drawing away students and fees. To make things worse, the director, Medicus, demanded special privileges and freedoms for his new professors. The elector and his officials, who seemed to ignore and disdain them at every turn, had shown special favor to the new institution. Now, as a final insult, they had to sit in the attractive hall of the fashionable new school and listen to the insults of some enthusiastic Pietist.[87]

All the while, supported by the "bravos!" of the audience and the approval of the assembled luminaries, Jung-Stilling sang the praises of academic cameralism and its heroes. He lauded Dithmar in Frankfurt an der Oder, Justi in Vienna and Göttingen, Zincke in Leipzig. He singled out Sonnenfels, Beckmann, and Schreber as the most important cameralists of their generation, and he placed Lautern's own Georg Succow among them. Jung-Stilling's emphasis is noteworthy. Beckmann, Schreber, and Succow were, after all, known for their command of the *Grund-* and *Quellenwissenschaften.* They were writers and teachers who knew, first and foremost, about chemistry, natural history, manufactures, trade, mining, and agriculture. And yet Jung-Stilling considered them, along with Sonnenfels, the foremost academic cameralists of their generation. The message was clear. The "basic sciences" and "resource sciences" had become full-fledged cameral sciences. It was the surest mark of mature cameralism. Subjects like technology, œconomic botany, and technical chemistry properly belonged to cameralist faculties and academies.

Medicus had not established the cameral academy to educate German political economists. Rather, Medicus, Schmid, Jung-Stilling, and Succow promised to educate *Kammerbedienten* who could manage state manufactories, mines, farms, forests, and universities. Equally important, academically trained cameralists would have to pick winners, deciding which state enterprises deserved support, and which had to be rejected. Jung-Stilling was clear about these in his 1786 speech. "Forestry, mining, agriculture, manufactures, and trade," he declared, "are the sources of all happiness for states and peoples—he who directs them must know them!—but all of these occupations rest on pure and applied mathematics, natural history, physics (*Naturlehre*), and chemistry."[88]

Lautern's commitment to an entire system of sciences led its proponents and practitioners to disdain books that seemed too general or theoretical.[89] In this regard, Smith's *Wealth of Nations* and the works of the French physiocrats seemed to have no substance. Jung-Stilling, for example, called physiocracy "a lovely girl, but unfortunately a virgin, who is incapable of making an honest man happy!" If physiocracy was seductive but barren, the benevolent young goddess of *Staatswirthschaft* promised riches for people and Kammer. Ironically, the cameral academy was lucrative for Karl Theodor and the Palatinate, but not in any of the ways that Jung-Stilling indicated. Rather, the cameral academy and the school of state administration that followed it generated money because they were fashionable, not because they were effective. In that sense, Medicus had chosen well when he appointed Jung-Stilling. He might not know how to run a farm, but, like Johann von Justi, he could sell almost anything, including the sciences.

CHAPTER SIX

Conclusion: Don't Believe Everything You Read

Schloss Friedenstein, the baroque palace that Pious Ernst built, is a good place to do research: on one side, in the east wing, the old research library; on the other side, in the west wing, the secret archive of Ernst the Pious. It is an ideal arrangement for anyone who wants to relate the cameralist literature to practices of fiscal administration. It is also an impressive physical space. Day by day, as I ate my lunch under Pious Ernst's statue and walked up the steep hill from town with his palace looming over me, the presence of the place—its vaults, walls, fortifications, chambers, and halls—had an impact. I started wondering about spaces and rooms, and about the Kammer; not the abstracted Kammer of fiscal-administrative history, or the idealized Kammer of the cameral sciences, but the *specific* room where Seckendorff met in council with the duke and his officials, and the *specific* chamber where the *Rentmeister* counted Pious Ernst's silver. Where was it? What did it look like? I have been unable to answer that simple question.[1] It is clear, however, that Gotha's Kammer was housed somewhere in the eastern half of the palace's main wing, near the church and the duke's living quarters.[2] That is where Duke Ernst's officials conducted the territory's fiscal business after 1654, when construction on that part of Friedenstein was complete.

In 1655 there was a large tabular chart hanging above the *Rentmeister*'s desk in the *Kammerstüblein,* or "little fiscal chamber" (figure 4).[3] It was probably there, hanging in the *Kammerstüblein,* when Seckendorff took charge of Gotha's Kammer in 1656. The chart was a distillation of Gotha's secret Kammer ordinance, which prescribed behavior for all members of the fiscal

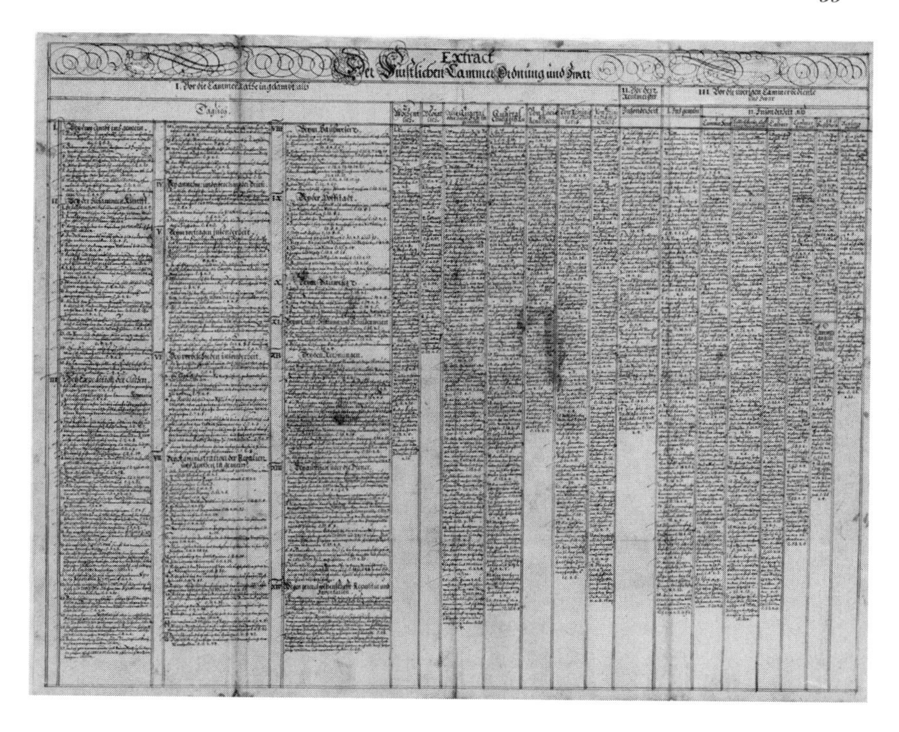
Extract

FIGURE 4 Table of duties from the Kammerstüblein (Thüringisches Staatsarchiv Gotha).

chamber. The table of duties was supposed to make the Kammer ordinance visible at a single glance, subdividing space and time to provide transparency about all tasks and expectations. Was it a form of social discipline, a cog in the "machinery of control that functioned like a microscope of conduct"?[4] Was it foreshadowing of the "statistical gaze" that eighteenth-century cameralists would use to enable and circumscribe the communication of experience?[5] That would be one way to read it.

Here is another way. The table of duties focused most of its attention on the *Kammerräte,* that is, the directors of the Kammer. It reminded them to vote according to rank, in orderly fashion, and meticulously to record their votes. It instructed them to avoid nastiness, verbosity, and bias, stressing that they should not laugh at other colleagues or make fun of them. It specified times for the collegial meetings of the Kammer: three hours every morning and three hours every afternoon, except Wednesdays and Sundays. It also specified a dizzying array of income sources that the *Kammerräte* had to manage, among them mines, beer, taxes, saltworks, streets, tolls, hunting, forests, streams, fish ponds, mills, and leaseholds.

It all sounds very disciplined, very demanding. But there is one small problem: there was no Kammer collegium. During Seckendorff's time in Gotha there was usually only one *Kammerrat;* Pious Ernst was so short-handed that the *Rentmeister* frequently had to assume control of the Kammer, something not even contemplated by the ordinance.[6] There were no meetings, no votes, no timetables, and no colleagues. The table of duties, and the secret ordinance upon which it was based, corresponded mostly to the fantasies of its author.[7]

I do not want to suggest that the ordinance, or the table of duties distilled from it, was unimportant. Ernst was concerned about the Kammer ordinance from the very beginning. After he became duke of Sachsen-Gotha, in 1640, he immediately collected ordinances from his brothers in Eisenach and Weimar, placing special emphasis on the Kammer. As a separate administrative entity, the Kammer was still a relative novelty in the mid-seventeenth century, and none of the Kammer ordinances collected by Ernst and his officials were more than a decade old.

There was much confusion about the relative spheres of Kammer and chancellery, of justice and fiscal administration. The 1636 Kammer ordinance for Sachsen-Eisenach, for example, complained at length about the confusion between "legal affairs" and "fiscal affairs."[8] When Christoph von Hagen came to Gotha from Sachsen-Eisenach as Duke Ernst's first *Kammerrat,* his immediate challenge involved carving out a separate sphere of administrative activity for the Kammer. Fifteen years later, in 1656, when Seckendorff replaced Hagen, there was still much "confusion" surrounding the Kammer and its proper sphere of activity.[9] The Kammer ordinance and Seckendorff's cameralist masterpiece, the *Teutscher Fürsten Stat,* appeared in the same year. They were not unrelated. The ordinance imagined the Kammer as it should be, not as it was. Similarly, Seckendorff's *Fürsten Stat* imagined Gotha as the well-ordered police state of Pious Ernst's dreams.

Despite these similarities, the differences between the Kammer ordinance and Seckendorff's *Fürsten Stat* were profound. One was a *secret* document, meant only for the members of Ernst's fiscal chamber; the other was a *public* document, meant to portray Gotha as a model principality. In this respect the *Fürsten Stat* resembled Ernst's *Landesordnungen,* his famous published ordinances about health and education.[10] But the *Fürsten Stat* was different from those published edicts in one key respect: it professed to sketch the intimate sphere of the prince. In that regard, it was altogether different from the Kammer ordinance, which, like everything else involving the Kammer and its officials, remained shrouded in secrecy. Every *Kammerrat* swore a lifelong oath of silence, deliberations were (officially) only allowed in designated "fiscal chambers," and all records relating to the Kammer were to be carefully guarded in the secret archive: "Nobody, whoever he be, whether he has an interest or not, shall get verbal reports or be allowed to read our reports, without the express consent of our *Kammerräte.*"[11]

Reading through the endless, monotonous records of the Kammer today—lists, inventories, account books, receipts, deeds—it can be hard to understand what all the fuss was about. But if you compare Ernst's Kammer directives to his published ordinances, the need for secrecy becomes clear. The documents in Gotha's secret archive were unequivocal about the purpose of the Kammer: one had to secure existing sources of revenue and find new ones by whatever means necessary. Police ordinances and cameralist texts might dwell on the "common good" and the "general welfare," on how the interests of the wise prince coincided with the interests of his subjects. But in the secret sphere of the Kammer, where it was a matter of filling the duke's treasury with silver, there was no time for that. Seckendorff, in keeping with the oath of secrecy, went to his grave with that knowledge. His private letter of resignation, which sat unread in the Seckendorff family archive for three hundred years, suggests that the secrecy and dishonesty demanded by the Kammer had finally overwhelmed him.[12]

SECRECY AND PUBLICITY

This has been a tale of two discourses: the public discourse of the cameral sciences and the secret discourse of the Kammer. I have tried to consider how they were related. The cameral sciences did not simply reflect the language of the Kammer, nor were they wholly unrelated to it. Instead, the discourses of Kammer and cameral science were distinct but interrelated, separate but interdependent. The cameral sciences lived parasitically from the status and secrecy of the Kammer. That is what made them so popular, elegant, and fashionable. By the same token, the Kammer benefited from the image fostered by the cameral sciences. Fiscal officials, with their taxes, imposts, and duties, generated plenty of hatred through the years, and with good reason. By providing a different image for the Kammer and its members—honest, intelligent, well-meaning—the cameral sciences provided an invaluable service. They *created* the good cameralist. Without the cameral sciences, that selfless state servant of eighteenth-century reveries might have remained the vicious bloodsucker of seventeenth-century nightmares.

So what *was* the relationship between the cameral sciences and administrative practice? Cameralist textbooks cannot resolve this question, no matter how many of them you read. Keith Tribe was right about that.[13] I have turned to different sources in search of answers, and these sources have corroded the boundaries of the thing I thought I knew, forcing me to reconsider it entirely. The cameral sciences were *strategic.* By painting idealized pictures of the fiscal-police state, cameralist texts served at the same time to promote it. As promoters of the well-ordered police state, cameralists served their own private interests, securing positions in territorial Kammern and German universities.

Does it matter? If we assume that cameralism was nothing more than a peculiar, early modern German economic doctrine, perhaps not. But the cameral sciences were not so limited as the secondary literature would have us believe. They were not merely "economic" in a modern sense; rather, cameral science, at once natural science and economic doctrine, political science and technological discipline, flowed freely across the disciplinary boundaries erected by later generations. Cameralists were not early modern analogues of economists or political scientists; they were equally chemists and foresters, mineralogists and technologists.[14] More than that, academic cameralists saw themselves as university-educated professionals, like physicians or lawyers. Throughout the eighteenth century, they made inroads at academies and universities, insinuating themselves and their professional vision into the very structure of the sciences. Their impact on natural philosophy and natural history was thus much greater than has been recognized.[15] In Göttingen, Curator Münchhausen helped put a crypto-cameralist faculty in place, complete with auxiliary sciences and œconomic gardens. Farther south, in Ingolstadt, Friedrich Casimir Medicus's "resource sciences" located the heart of cameralism in natural knowledge.[16] The cameral sciences were natural sciences just as much as they were economic sciences. The well-ordered police states they promoted were as much visions of nature as they were political-economic constructs.

In their public lives—published texts and academic lectures—cameralists remade disordered worlds into well-ordered police states; in their secret lives they confronted a universe of failing mines and cheating officials, of recalcitrant nature and deceitful humanity. The satirists had it right: cameralists were necessarily dishonest. While purporting to speak publicly about the sovereign's most secret affairs, they manufactured utopian principalities filled with pious subjects, healthy animals, thriving crops, and honest officials.

SKEPTICISM AND DISORDER

We recreate the past from the vestiges of fossilized language: new books, old books, manuscripts. Some of us "read" pictures and objects for access to forgotten worlds. There is no trouble about that. True scholars, we like to think, immerse themselves in time past. They live there. It is a salutary myth for the historical profession, a noble lie that protects history from unpleasant incursions by sociologists, economists, and other present-minded people. Most historians really do treat the past as if it has integrity and deserves respect. It is what separates us from other disciplines. We may be right to do so, or maybe it is a fetish.

Historians, as a professional group, can still be unreflective about how we excavate the past.[17] That has changed some in recent decades, but we still make confident pronouncements about how it all was. There are many things, obvious things, that I do not know. I can tell you a lot about Foucauldian panopticism, Weberian charisma, or Elias's civilizing process. But don't ask me whether the *Rentmeister*'s lockbox had silver in it, or who had the keys. I don't know. This is not just ritual abasement. As gatekeepers of the past, historians can be bad about acknowledging the vast spaces of our ignorance; we should do it more often.

This is a skeptical book, a call to reexamine many of the things we thought we knew.[18] I feel much less sure of myself today than I did a decade ago, when I embarked on this project. I am not so sure anymore that the well-ordered police state was well ordered; not convinced that the disciplinary dreams of eighteenth-century administrators correlated with disciplined subject populations; not sure what the "descriptive sciences" actually described; and not convinced that there was any connection between eighteenth-century science and economic development. Like the practitioners of *Alltagsgeschichte,* I am skeptical about the implicit teleology of large explanatory structures: "modernization," "industrialization," "bureacratization," and so forth. Like them, I embrace contingency and find inspiration in the micro-scale. But I do not have their patience. Gotha might have been enough; it could have been my Neckarhausen or Laichingen.[19] The truth is, though, that I wanted to compare and generalize even as I benefited from the insights of small-scale analysis.[20] So I have sampled the empire's territories like a quality controller, testing the occasional silver mine or pine forest for rotten narratives. This may not produce certainties, but it certainly does raise questions.

One thing is clear: we need to be more careful about sources and evidence. I have come to appreciate the seductive power of large, often unacknowledged, narrative structures. Helmholtz long ago demonstrated how human beings navigate the unknown by falling back on the familiar. Should it come as any surprise, then, that I once saw evidence of the Foucauldian gaze in Gottlob Christian Happe's desperate rantings about surveillance, or that I discovered the relentless beat of social disciplining in vain attempts to regulate the ungovernable forests of the Harz? I do not want to suggest that I have achieved some kind of Baconian objectivity, exorcizing my own idols of cave and tribe. I am not that pure. Still, every time and place has its Aristotle, the dominant philosophical and ideological lens through which it perceives the world. Many German historians, and historians of science too, have been conditioned to see signs of creeping order, discipline, and surveillance around every corner. The recently codified Foucauldian-Weberian disciplinary society has granted abstractions like "disciplinization" and "bureaucratization" a life of their own, creating a historical world in which these forces work insidiously, like gravity, to form states and shape individuals.[21] The cameral and police sciences have assumed an important place in such narratives, linking specific disciplinary mechanisms to the rise of the social sciences.[22] These new disciplinary syntheses connect in uncomfortable ways with national stereotypes. There is something viscerally persuasive, for example, about methodical Prussians ordering the human and natural worlds.[23] German historians may have abandoned a history that identifies eighteenth-century Romantics with proto-Nazis, but the Third Reich continues to cast its long shadow back across the old Reich in subtle and unexpected ways.[24] By the same token, we all love a free and rational Englishman harnessing science to fuel the Industrial Revolution.[25] National stereotypes die hard.

The historiography of cameralism owes much to national stereotypes. For Albion Small, writing shortly before World War I, cameralists were the shock troops of German modernity. Facing a population of infants—"infants in knowledge, infants in experience, infants in feeling, infants in judgment about the conduct of life"—German princes tried to arouse the "dormant powers" of these backward people.

> The method by which the German leaders undertook this task was something like the method by which a levy of raw recruits is made over into a regiment of disciplined soldiers. The Germans were divided up

> into some hundreds of squads, each controlled by a territorial prince who was within limits absolute in his own land. . . . Populations largely of peasants, and the remainder mostly artisans who had been incubated in the quasi-communistic guild organizations, and had never learned to walk alone, populations politically and economically in their swaddling clothes, and needing, first, nursery care, then tutors and governors to bring them to maturity—this was the situation in which that paternalism culminated which Americans have been taught to despise.[26]

Small's "regiment of disciplined soldiers" calls to mind Foucault's celebrated riff on the Prussian army of Frederick II. Foucault used Frederick's infantry regulations to make his points about disciplined bodies and disaggregated time. "The more time is broken down, the more its subdivisions multiply, the better one disarticulates it by deploying its internal elements under a gaze that supervises them, the more one can accelerate an operation, or at least regulate it according to an optimum speed."[27] Max Weber, too, invoked time and discipline, although with more emphasis on religion, as his Calvinists stamped their spiritual mark on the matter of economy and society.[28] Gerhard Oestreich's neostoics, with their endless police ordinances and resolute social disciplining, did their part as well, training the common people to lead well-ordered lives.[29] There were also all those well-mannered French at Versailles who helped drive Norbert Elias's civilizing process.[30] The rising crescendo of early modern disciplining has reached its climax in the so-called "disciplinary revolution"—a shot of Foucault and Weber, together with a splash of Elias and Oestreich.[31] Albion Small's disciplined cameralists seem to be everywhere now, spreading order, surveillance, incarceration, and bureaucracy throughout early modern Europe.

I guess I have my doubts. The cameralists I know—Becher, Seckendorff, Justi, Beckmann, Schreber, and Jung-Stilling among them—were not very well behaved. They lied, connived, cheated, and embezzled. But they also painted beautiful pictures of well-ordered police states, complete with thriving populations, useful sciences, flourishing manufactures, industrious farmers, and disciplined state officials blessed with knowledge and integrity. They did it because their lives depended on it. Justi floated from country to country, hawking one project after another and trying to live from his writings. Others, like Beckmann and Jung-Stilling, landed positions as academic cameralists. Their professional survival depended on the ability to attract wealthy students, and so they became dedicated promoters of the cameral sciences.

In territory after territory, from tiny Gotha to middling Hannover and sprawling Prussia, cameralists sketched happy images of well-ordered possible worlds for their lords and patrons. Savvy ministers like Münchhausen and Heynitz, who understood the true utility of cameralist treatises, used these men to promote their towns and attract wealthy students. Others, however, were taken in by all the fancy talk—even Frederick the Great fell for Justi's sales pitches. Not surprisingly, then, many historians have been bamboozled too, taking all those thriving mines, healthy forests, disciplined officials, and fat cows at face value. It is time to reconsider.

We have covered a lot of ground, from the *Thüringerwald* to the silver states of Saxony and Hannover, from the Prussian bogs around Küstrin to the Palatine forests near Lautern. In each of these places I have tried to relate the secret discourse of the Kammer to the public discourse of the cameral sciences. Patterns have emerged, and now it is time for definitions. Cameralism was the public face of secret things; cameralists were publicists for the Kammer. Cameralism did not simply reflect administrative practice in well-disciplined German principalities, nor was it wholly unrelated to fiscal administration. Rather, cameralists *created* the well-ordered police state through their ordinances, books, and treatises. But behind these well-ordered visions lurked a disordered world of fear and frustration. For all his profound administrative maxims and chemical principles, Justi failed to control the human and natural worlds, ultimately sinking under the weight of bad iron ore and confused account books. Despite his brilliance at systematizing the "resource sciences," Jung-Stilling was defeated by bad manure and scrawny cows. Seckendorff, who sang the praises of Pious Ernst and his model principality, was plagued by the disorder and dishonesty of Gotha's Kammer.

Cameralists were fiscal propagandists. They argued that a well-organized structure of human and natural sciences—police science, economy, chemistry, forestry, mineralogy, and so on—would yield prosperity; they claimed that skilled cameralists could wring new revenues out of mines, forests, and domains by harnessing that knowledge; they presented themselves as servants of the general welfare. In the secret space of the Kammer, however, these same cameralists focused resolutely on the interests of the prince and his treasure, developing new techniques to fleece the people. As the public representation of secret affairs, the cameral sciences were essentially dishonest. Our anonymous satirist, Maria Machiavel, who depicted the cameral sciences as ideological cover for a bloodsucking Kammer, saw it that way. Moreover, such public representations of secret affairs, whether in the form of forest edicts or systematic treatises on finance, often functioned more as good publicity than as tools of discipline. Happe, with his despair about pointless police ordinances and his fantasies about panoptic cabinets, made that point. Finally, though academic cameralism posed as knowledge that could change the world, it was actually useful in a different sense: once the cameral sciences became fashionable, they drew wealthy students to places like Göttingen, Freiberg, and Lautern. Those students, like Barkhausen's fictional young cameralist, might learn nothing useful at university; from the standpoint of the Kammer, however, the cameral sciences demonstrated their utility by attracting wealthy academic customers to backward little towns.

Cameralists were not what they seemed to be. Justi continues to con us today, much as he seduced Frederick the Great more than two centuries ago. We still habitually conflate Seckendorff's model police state with an actually existing principality. And Jung-Stilling's romantic pragmatism, with its detailed descriptions and down-to-earth language, still convinces us that his utopian pigs and cows, fat and healthy, corresponded to real creatures. We have trusted them all too much. And that is no small thing, because cameralism, broadly construed, constituted a sprawling body of literature that spilled across the disciplinary boundaries erected by generations of historians, sociologists, economists, and political scientists after the fact.[32] Cameralists may be extinct, but their well-ordered police states live on in our histories; their fantasies about controlling man and nature have become our evidence for the disciplined past.

And so we return to where we started, to Süss Oppenheimer, hanging high above Stuttgart in his scarlet-red cage. Süss did not write textbooks about proper budgeting; he did not reveal the fundamental principles of state administration; he did not lecture about the general welfare. Still, he was called a "cameralist." Why? Because Süss Oppenheimer was a scapegoat of the Kammer. The people of Stuttgart, mixing older currents of anti-Semitism with fresh resentments over taxes, vented their rage on him. His case was not atypical. Every cameralist was a potential scapegoat, existing in that tenuous space between sovereign displeasure and public retribution. He might, like Süss, be sacrificed to the people; he might, like Justi, rot in prison for displeasing the prince.

Danger focuses the mind. Cameralists had to be skillful rhetoricians because their lives and livelihoods depended on it. Later generations mistook bravado for confidence. But there was nothing secure about the itinerant Johann von Justi, living day-to-day from his writings, hounded by creditors and his wife, pursued by European monarchs, thwarted by rebellious iron workers, confounded by nature herself. Only in the well-ordered worlds of his imagination was Justi sovereign. He escaped there by candlelight, safe from the tortures of everyday failure and frustration. Jung-Stilling and Seckendorff did the same thing.

I have argued that German cameralists were important. You are right to distrust me, because this is a book about German cameralists, and I want you to read it. For the same reason, you cannot trust many of the sources that sustain our grand narratives about science, government, and economy in the Enlightenment. Our cameralists had every reason to suggest that systematic knowledge, carefully cultivated by good princes and their officials, would benefit the general welfare. They had every reason, that is, to connect the sciences with discipline, prosperity, and material progress. It is strange that we have believed them.

APPENDIX ONE

Average Annual Silver Production in Central Europe (kg.), 1545–1800

Years	Electoral Saxony	Hannover/ Braunschw.-Wolfenbüttl.	Austrian Habsburg lands	Other	Total
1545–1560	13,150	3,400	30,000	3,300	49,850
1561–1580	9,000	3,500	23,500	2,500	38,500
1581–1600	7,550	4,500	17,000	2,250	31,300
1601–1620	5,100	3,550	11,000	1,750	21,400
1621–1640	3,100	1,900	8,000	1,000	14,100
1641–1660	2,600	2,800	8,000	1,100	14,500
1661–1680	2,950	2,900	10,000	1,150	17,000
1681–1700	3,800	5,850	10,000	1,750	21,400
1701–1720	4,800	8,050	10,000	2,500	25,350
1721–1740	6,500	15,200	12,500	3,500	37,700
1741–1760	6,800	10,850	24,000	3,500	45,150
1761–1780	7,200	7,900	24,000	3,000	42,100
1781–1800	12,100	7,800	26,000	4,000	49,900

Source: Based on tables in Adolf Soetbeer, *Edelmetall-Produktion und Werthverhältnis zwischen Gold und Silber seit der Entdeckung Amerika's bis zur Gegenwart,* 107–8. Soetbeer's estimates are a good general approximation, but not perfect. For detailed commentary on and criticism of Soetbeer's estimates, see Nef, "Silver Production in Central Europe, 1450–1618." Specialized studies provide more reliable figures for individual regions. Christoph Bartels's study of mining in the upper Harz, for example, includes extensive estimates of annual silver production (Bartels, *Erzbergbau im Oberharz*, 726–31).

Note: The Austrian Habsburg lands include mining regions in Austria, Bohemia, Hungary, Tirol, and Transylvania.

APPENDIX TWO

Acquisition History of Selected Mining Books in Göttingen

Year and place of publication	Author	Short title	When acquired by Göttingen	How acquired by Göttingen
1700, Dresden	Rößler, B. Edited by J. Goldberg.	*Hellpolierter Bergbauspiegel*	Part of older collection	Bülow's library
1725, Leipzig	Henckel, J. F.	*Pyritologia oder Kieß-Historie*	Winter 1749	—
1749, Schneeberg	Beyer, A.	*Gründlicher Unterricht vom Bergbau*	1st ed.: 1749 New ed. (1785): July 1785	1st ed.: — New ed.: Dieterich
1749, Dresden	Oppel, F. W. v.	*Anleitung zur Markscheidekunst*	Winter 1749	—
1750, Leipzig	Gellert, C. E.	*Anfangsgründe zur metallurgischen Chimie*	1st ed.: Summer 1751 New ed.(1756): July, 1776	1st ed.: — 2nd ed.: Dan. Fr. Kübler
1755, Leipzig	Gellert, C. E.	*Anfangsgründe zur Probierkunst*	1st ed.: Spring 1756 2nd ed. (1772): 1854	1st ed.: — 2nd ed.: Gift from "Hof-buchhandler Hahn" in Hannover

Year and place of publication	Author	Short title	When acquired by Göttingen	How acquired by Göttingen
1763, Braunschweig	Calvör, H.	*Historisch-chronologische und theoretische Beschreibung des Maschinenwesens*	November 1763	Arrived in the mail on 28 Nov. 1763
1769, Freiberg	Kern, J. G. (anon.). Edited by F. W. v. Oppel.	*Bericht vom Bergbau*	July 1769	Gift to Göttingen library from Oberberg-kommissar Heynitz
1770, Prague	Peithner, J. T. A.	*Erste Gründe der Bergwerks-wissenschaften*	April 1772	Sent as part of large shipment from Vienna
1771, Prague	Poda, N. Edited by J. v. Born.	*Kurzgefaßte Beschreibung der bei dem Bergbau zu Schemnitz . . . errichteten Maschinen*	December 1771	Dieterich
1773, Vienna	Delius, C. T.	*Anleitung zu der Bergbaukunst*	August, 1774	Dieterich
1773, Dresden	Poda, N. Edited by D. Breitenheim.	*Akademische Vorlesungen über die zu Schemnitz . . . errichteten Pferdegöpel*	November 1773	Vandenhoeck

APPENDIX THREE

Friedrich Casimir Medicus's Unpublished Proposal for a Faculty of State Administration at the University of Ingolstadt

NOTE ON THE FOLLOWING PLAN

At universities and also in councils of state, the importance of the science that has until now gone under the name of cameral science is generally acknowledged. But in both cases it rests on merely fragmentary knowledge; that is the reason why this science, so long the dream of states, has had so little influence on the welfare of the state. Not only has it had little influence on the welfare of the state, but it has even produced the opposite effect, since instruction has until now generally bred unskilled and unlucky projectors, a plague of state much more dangerous than the former ignorance.

True patriots must also assume that correct knowledge is presented at universities; when this is not possible, they would rather oppose every lecture strenuously. For if they cannot do any good, their duty is at least to do no harm.

Since one gives me the surpassing honor to know my thoughts about the course of study already established at Ingolstadt, I cannot hide that I wish it was still as it was a few years ago, namely, that this science was not being taught there at all. The Herren Schlögel and Moshammer[1] only lecture on parts of the whole. The outcome of their efforts can therefore only be either useless or harmful.

It is a generally recognized truth that for every subject there must be a foundational system. Nowhere is this more necessary than in state administration (*Staatswirthschaft*), because without a well thought out system, lord and land will be ruined.

The truth of this saying is so generally recognized that other princes of the Empire, who now hope to introduce the teaching of this science, have preferred the systematic approach ever since your electoral majesty established the academy in Lautern. Examples are the newly established faculties in Gießen and Mainz. In Ingolstadt alone did one remain with the former flawed and harmful approach.

Since one now however has a patriotic wish to kill this error in its infancy, and to establish a thorough course of instruction, it is necessary to establish a separate faculty there.

Names are supposed to be meaningless. But experience convinces me of the contrary. In Lautern this academy was called "cameralist." And since one interprets "*Kammerale*" as finance, so one thought that it was only designed to instruct high fiscal officials *(Hofkammerräte),* which is an extremely dangerous opinion. The fiscal official and the state administrator (*Staatswirth*) are two very different people. The former thinks only about the increase and maintenance of the princely income, and thus too often ignores the welfare of the land. The fiscal official must therefore be subject to the state administrator and must be guided by him. This merely inapt designation of the academy has generated a completely mistaken impression of it with the public; this has caused me much trouble, and the refutation of it, which has not even succeeded, has also cost me much effort.

In Gießen, therefore, one eventually called this faculty "œconomic." But then the public believed that one was only lecturing about agricultural things and intended to educate farmers. In Mainz one eventually connected both names, and called the faculty the "œconomic-cameralist." The effects were like those mentioned above. I consider it very important, therefore, to give the new faculty a name that denotes its large and important range, while indicating to both the knowledgeable and the ignorant its proper domain. In short, one should call it the

Faculty of State Administration (*Staatswirthschaftliche Facultät*). I wholeheartedly believe that only this name will effectively eliminate those difficulties which caused us the greatest trouble in Lautern.

When the academy in Lautern was founded I believed, and still do, that it had to be established in some isolated place. Your electoral majesty has thereby become a founder of this science in Germany, initiating an epoch that will stand as an eternal monument for posterity. I also believe that it would be much too early to move the academy to Heidelberg, because the science has not yet developed sufficient roots, and I worry that it could be destroyed through incorporation by those who secretly oppose and envy it. But I do not therefore advocate dividing this curriculum among other universities and transporting it to those places. This curriculum got its form in Lautern. In Lautern it must find its continuation and its consistency. And if the academy stays in Lautern, then it will be easy to introduce and maintain this new curriculum at other universities. And I hazard to guarantee that the introduction of lectures on the science of state administration in Ingolstadt will encounter almost no difficulty if our academy remains in Lautern. But [I fear] that it will have trouble succeeding in both places if the school is transplanted to Heidelberg. The envy so sadly typical of scholars poses a mighty hindrance, which will be checked if they see that the state is prepared to change the situation at any moment if it [sees] that one wants to hinder the truly useful through cabals.

As much as I wish that his electoral highness will keep the academy in Lautern, and establish it there, so much do I wish that this science will be taught nowhere else in Bavaria except Ingolstadt.

NONBINDING PROPOSAL FOR A FACULTY OF STATE SCIENCES AT THE UNIVERSITY OF INGOLSTADT

Where mastery of a science is concerned, the most important thing that the academy's director must ensure is a systematic course of study; pupils must learn principles before being introduced to practice. These maxims, though well known, have been entirely ignored by those who introduce this course of study, so that they present only certain parts of the whole as public lectures or, even worse, they instruct their pupils to perform things about which they have absolutely no idea.

Our illustrious elector is the first prince in Germany who has forcibly resisted this, and your highness was the first to introduce the systematic instruction of so-called cameral science in your states, something that will always constitute an important aspect of your highness's biography. Ten years of reflection, and the direction of the Lautern Academy, have acquainted me with this course of study, and with everything that is necessary for it; based on my experiences, I would like to outline my thoughts about the creation of this new faculty.

The science rests

1. on foundational doctrine (*Grundlehre*),
2. on the doctrine of resources (*Quellenlehre*),
3. and on state administration proper.

1. Foundational teachings consist of natural history, physics, chemistry, and mathematics.

Natural History acquaints the future state manager with those things that he will, with time, apply for the profit of the land. Agriculture, forestry, and all the business of society rest on it. Its teaching is indispensable, and so I wish that a teacher be appointed for just this purpose, for which position I can think of no one who would dedicate himself to this office with more passion and more profound knowledge than Herr Schranck of Burghausen.

Naturlehre (physics) is taught by Herr Professor Staiglehner, a very thoroughly learned man, and chemistry by Herr Rouseau, who has also made himself very famous in this field. But because Herr Rouseau is a man of years, Herr Schranck might be encouraged to study chemistry with zeal under Herr Rouseau, so that after his death this professorship could be combined with the one in natural history. For both sciences have the closest connection, and it is impossible, for example, that someone who is not a chemist can be a thoroughgoing mineralogist.

Mathematics is the most indispensable science. On the one hand, pure mathematics is the best logic, which teaches the future state administrator how to think correctly. On the other hand, applied mathematics is indispensable in all factories, mines, mills, etc., and most of all for architecture. The large range of courses demands its *own* teacher, and the importance of the subject demands a very *thorough* teacher. I would therefore like a new teacher to be appointed for this purpose, someone who is known for his profound knowledge.

The foundational sciences that have been outlined here are the true basis upon which all the knowledge of a future state administrator rests, and without which he can never make a sure step forward. One must therefore guarantee that no pupil who has failed eagerly to study these sciences is allowed to proceed to the resource sciences.

2. The *resource* teachings are those sciences that encompass the wealth of the state. These are agriculture, forest science, and mining, or the extraction of raw products. In addition, there are courses about artisans and manufactories, or the processing of raw products. Finally comes the science of trade, or

marketing by exchange, or sale of raw and processed materials. Only *one teacher* may teach all of these Resource Sciences, so that one part is not improperly raised above the others, after the fashion of the French œconomists.[2] The state administrator must treat all of them equally, and must value and honor each one as he does the others. For this purpose a specially designated teacher must be appointed. On the basis of long experience I propose the young Herr Semmer, who has studied this science in Lautern with exceptional diligence, and who will bring true honor to the professorship.[3]

3. *Actual state administration* (*Staatswirthschaft*) has three parts. *Police science* shows how all the occupations of civic life, which I sketched above under the resource sciences, must be directed so that each of them flourishes in the state. *Financial science* teaches how the state's income, which is necessary to sustain the court, as well as the civil service and the military, can be collected and profitably employed without harming the [objects of the] resource sciences.

Finally, true state administration teaches the art of discovering the laws that bring these aforementioned sciences to life and make them effective in each land. Herr Moshammer will take over this professorship, which is already provided for.

Not only pupils but even their fathers believe that this science alone is necessary; they thus hope to study only this one and to avoid the foundational and resource sciences. One must therefore ensure that only those who have already taken the other designated sciences be allowed take courses in this one.

Since incalculably much depends on how these courses are presented and which textbooks are used, the highest state Curatel in university affairs should order the newly established faculty humbly to submit for approval the instruction plan for the coming semester, together with the textbooks to be used, two months before the beginning of every term.

Certain aids are necessary for this faculty:

1. The university library must acquire the most important works necessary for these sciences.
2. It is important to check that the plants in the botanical garden are planted in accordance with Succow's œconomic botany, so that the teacher of natural history can indicate them.
3. There should be a mineral cabinet, not for show but only for teaching; it need therefore not be large but only has to be instructive.

4. If there is no chemical laboratory, then one needs to be built with public funds.

A great secret of every high university Curatel involves linking the interests of the professor to the welfare of the university, which contributes mightily to the success of a university. Each teacher should thus have the right to charge for his courses. This awakens the zeal of the teacher, and creates praiseworthy emulation.

The highest Curatel in university affairs has, it is true, determined that all students should contribute a certain designated sum for the courses of the four faculties. Though I would like this rule to stay in place for these faculties, an exception should be made for the new faculty alone. This exception would contribute incalculably much to the success of the new faculty, as I will humbly demonstrate later in the Special Order.[4] I would also like each course offered by this new faculty to cost five florins per semester. Moreover, five hundred florins is a very respectable salary that will make each professor manage his office with enthusiasm and devotion.

According to this proposal (not considering the salaries of the Herren Staiglehner, Rouseau, and Moshammer, who were appointed long ago), the establishment of the new faculty might cost

1. for three new teachers at five hundred florins each: fifteen hundred florins,
2. for acquisition of materials: five hundred florins.

If the latter five hundred florins are used wisely each year, then in a short time the necessary materials can be acquired; little by little it can grow in such a way that in a few years there will be a considerable stock, which will make this faculty shine.

A possible suggestion would involve moving the economic society from Burghausen to Ingolstadt, including members of the new faculty in it, and incorporating its fund into the faculty budget. Regarding this, it is worth noting that Herr von Hartmann would have to remain vice president of the society, though the meetings would be held in Ingolstadt where they could assume the proper form.

This constitutes, then, the proposal and the main plan. Should it be approved, each teacher would need to receive his own orders, and the details would have to be arranged to create a perfect whole.

NOTES

CHAPTER ONE

1. On the execution, see Haasis, *Joseph Süß Oppenheimer,* 434–49; also Feuchtwanger, *Jud Süss,* 599–611.

2. Cf. Haasis, *Joseph Süß Oppenheimer,* 405–32; and "Jud Süss," at the Web site of Shoa.de: http://www.shoa.de/jud_suess_person.html.

3. [Machiavel], *Der volkommene Kameraliste,* 60. Maria Machiavel is the pseudonymous author of this satire. I have not been able to determine the identity of the author. Another edition appeared in Köln in 1765.

4. Recktenwald, "Cameralism," 1:313–14.

5. For a more extensive review of this issue, see Wakefield, "Books, Bureaus, and the Historiography of Cameralism." On cameralism as a university discipline, see Stieda, *Universitätswissenschaft;* and Tribe, *Governing Economy.* On cameralism as baroque science, mode of thinking, or system of principles, see Walker, *German Home Towns;* Outram, *The Enlightenment,* 102–4; Brückner, *Staatswissenschaften, Kameralismus und Naturrecht.* A few authors have acknowledged administrative practice more explicitly. See, for example, Schumpeter, *History of Economic Analysis,* 159–60; and Lindenfeld, *Practical Imagination,* 14. For a systematic effort to link the cameral sciences with practices of administration, see Osterloh, *Joseph von Sonnenfels.*

6. For opposing views on this, see Small, *Cameralists* and Tribe, *Governing Economy.*

7. See Roscher, *Geschichte der Nationalökonomik,* 228–31; Schmoller, *The Mercantile System;* Sommer, *Die Österreichischen Kameralisten;* Schmidt-Bielicke, "Der Autarkiegedanke im Merkantilismus"; and Focke, "Die Lehrmeinungen der Kameralisten über den Handel."

8. Small, *Cameralists,* 591.

9. Ibid., xiii.

10. Zielenziger, *Kameralisten*, 85–110. See also Nielsen, *Die Entstehung der deutschen Kameralwissenschaft*, 1–10; and Dittrich, *Kameralisten*, 14–15.

11. Zielenziger's influence extended to Anton F. Napp-Zinn and others, who argued that cameralism reflected the administrative apparatus of the German territorial state. See Napp-Zinn, *Johann Friedrich Pfeiffer*, 10; Dittrich, *Kameralisten*, 14–15; Sommer, *Österreichischen Kameralisten*, 43–56, 62–105.

12. Small, "What Is a Sociologist?," 470–71.

13. Small, *Cameralists*, 6.

14. Zielenziger adopted Small's term *fiscalists* to describe those servants of the Kammer who were not cameralists proper, that is, writers. See Zielenziger, *Kameralisten*, 86–87.

15. Ibid., 99–101.

16. Tribe, *Governing Economy*, 10–17.

17. Few seem to have recognized the radical nature of Tribe's argument, or that it is essentially incompatible with older approaches. Instead, it has often been folded together with older works, as if it was an extension of them. See, for example, Sheehan, *German History*, 193–94.

18. There were several variants of the term, stretching from the seventeenth into the eighteenth century, many of which make clear the original connection between the cameralist and the Kammer, or fiscal chamber. These included *Cameralist*, *Cammeralist*, *Kameralist*, and *Kammeralist*.

19. See Becher, *Politische Discurs*, 889–908; Schröder, *Fürstliche Schatz- und Rentkammer*, 11–13, 27–30. By the eighteenth century, the term *Kammer* had become the generally recognized shorthand for the treasury or fiscal chamber of the prince; during the seventeenth century, however, authors still modified the term *Kammer* in various ways to clarify that they were speaking about the fiscal chamber and not, say, the bedchamber. Becher, for example, wrote specifically about the "*Finanz-Kammer*" and Schröder about the "*Rent-Kammer.*"

20. Schröder, *Fürstliche Schatz- und Rentkammer*, 11–13, 27–30.

21. Ibid.

22. The following account draws on "Dr. Bechers Gutachten wegen rechter Bestellung einer Hoff- oder Finanz-Cammer," reprinted in Becher, *Politische Discurs*, 889–908.

23. In German: "den Unverstand, den Unfleiß, die Unordnung und die Untreu"; see Becher, *Politische Discurs*, 893.

24. Ibid.

25. See chapter 3.

26. Justi, *Abhandlung von den Mitteln die Erkenntniß in den Oeconomischen und Cameral-Wissenschaften*, 10–14.

27. Ibid., 1–19; and Justi, *Staatswirthschaft*, xxxi–xlii.

28. On the importance of "ordinary revenues" see Becher, *Politische Discurs*, 891.

29. [Machiavel], *Der volkommene Kameraliste.* J. C. E. Springer panned the work in the *Allgemeine deutsche Bibliothek*, claiming that it was derivative of Jonathan Swift's *Directions to Servants.* This criticism, while unfounded, is not surprising, because Springer himself was a published academic cameralist who had worked in Göttingen. See Springer, "Review of *Der volkommene Kameraliste,*" 296. See also Carpenter, *Dialogue,* 95.

30. Frederick II, King of Prussia, *Anti-Machiavel.*

31. [Machiavel], *Der volkommene Kameraliste,* 5.

32. Ibid., 7–8, 10. I translate the idiosyncratic *Das Kamerale* here as "cameralism," which is not ideal but seems to me the best available option. Justi, the only cameralist author specifically mentioned in the text, is cited several times.

33. Ibid., 8.

34. Ibid., 15–16.

35. Ibid., 17–18.

36. Ibid., 21.

37. Ibid., 22.

38. See Justi, *Staatswirthschaft*, preface.

39. [Justi], *Brühl.* There has been some controversy about the book's author, but the consensus has long been that Justi wrote it, which is confirmed by style and circumstance. The only serious recent challenge to this claim appears in Jürgen Luh's analysis of Brühl, "Vom Pagen zum Premierminister," where he asserts that because the book was "commissioned" by Frederick II, it could not have been written by Justi (121–22). But it was precisely in 1760 that Justi entered the service of Frederick II, which had long been his goal (cf. Frensdorff, "J. H. G. von Justi," 81). Justi may have lampooned Brühl to curry favor with Frederick (he often wrote books for that reason).

40. See Luh, "Vom Pagen zum Premierminister," 121–24.

41. [Justi], *Brühl*, 1:20–38.

42. Ibid., 1:20–38, 55–56, 127, 162–64. It is telling that Justi took special trouble to dismiss accounts of Brühl's education in Leipzig.

43. Justi, *Staatswirthschaft*, v.

44. There is no equivalent for *Wissenschaft* in English. What Justi means here is a body of structured knowledge that can be taught. The analogy to law or medicine is more apt here than, say, physics.

45. Justi, *Staatswirthschaft*, xxviii.

46. The author, Gottlob Christian Happe, was a jurist with close family ties to Thuringia. His father, Volkmar Happe, had served as secret councillor and premier minister in Weimar.

47. "G. C. v. Happe" to Friedrich II, 17 April 1717. The letter was enclosed in Happe's anonymously authored *Nichts Bessers, als die Accise, Wenn man nur will* (1717). I discovered it in the summer of 2004 in Schloss Friedenstein (Gotha). It has since been catalogued in the manuscript collection of the research library: FBGotha,

Handschriften, Chart. B 1918 II, "Happe, Gottlob Chr." The following account is based on Happe, *Accise*, 38–68.

48. Emphasis in original.

49. Cf. Wegert, "Contention with Civility," 360; Gross, "Temporality," 53–82; Ramati, "Harmony at a Distance," 437; Engelstein, "Combined Underdevelopment," 314, n. 11; Wolin, "Democracy and the Welfare State," 467–500. Broman, "Rethinking Professionalization," 857, n. 54; Gordin, "The Importation of Being Earnest," 1–31; Holquist, "Information Is the Alpha and Omega of Our Work," 415–50.

50. Raeff, "Well-Ordered Police State," 1226–1228. Raeff's claim that the German model was then taken up in Catherine II's Russia, while fascinating, is beyond the scope of my investigation. He expanded the argument in his 1983 book, *The Well-Ordered Police State.* Raeff's argument bears a family resemblance to Gerhard Oestreich's theses about social disciplining. Cf. Oestreich, *Neostoicism and the Early Modern State.*

51. Raeff, *Well-Ordered Police State,* 7–8.

52. Walker, "Rights and Functions," 235, n. 1.

53. See, for example, Linke, "Folklore," 135; and Breuilly, "Hamburg," 705.

54. Gross, "Temporality," 61; Linke, "Folklore," 135.

55. Though the literature here is enormous, the lead dogs have been Michel Foucault and Gerhard Oestreich. For a start, see Foucault, *Discipline and Punish* and "Governmentality"; Oestreich's *Neostoicism and the Early Modern State* is also a good introduction.

56. See Rau, *Kameralwissenschaft,* 2–10.

57. See Heß, *Geheimer Rat und Kabinett*, 1–161.

58. Hintze, "Hof- und Landesverwaltung," 138; Rosenthal, "Die Behördenorganisation Kaiser Ferdinands I.," 51–94.

59. Klinkenborg, "Die kurfürstliche Kammer," 220.

60. Oestreich, "Das persönliche Regiment," 220; Ohnsorge, "Zur Entstehung und Geschichte der Geheimen Kammerkanzlei," and "Zum Problem: Fürst und Verwaltung," 150–57.

61. Heß, *Geheimer Rat und Kabinett*, 1–3. Kurt Dülfer charted some variation across territories. See Dülfer, "Studien zur Organisation des fürstlichen Regierungssystems," 237.

62. See Troitzsch, *Ansätze*, 7–10; Small, *Cameralists*, xiii, 6; Lindenfeld, *Practical Imagination*, 2–4.

63. See Kraemer, "Der deutsche Kleinstaat"; Stahlschmidt, "Policey und Fürstenstaat." More recently, see Schilling, *Höfe und Allianzen*, 136; Vierhaus, *Staaten und Stände*, 99; and Münch, *Das Jahrhundert des Zwiespalts*, 127.

64. See Stahlschmidt, "Policey und Fürstenstaat." Seckendorff's job description is in the ThStAGotha, KK III, Nr. 8, "Kammerordnung."

65. Andreas Klinger has written an outstanding account of the "real" Fürstenstaat. On Gotha's finances, see his *Gothaer Fürstenstaat,* 186–211.

66. Ibid., 127.

67. Ibid., 192–94.

68. Seckendorff, *Lob-Rede,* especially verses 22–35.

69. The report on what happened, complete with toll receipts and letters from tree drivers, runs over one hundred pages. It shows just how far removed life in the everyday Kammer was from the happy visions of cameralist writers. See ThStAGotha, Kammer Immediate, Nr. 1201.

70. See the local forest commission reports contained in ThStAGotha, Kammersachen Insgemein, Nr. 1041, Nr. 1042; also Heß, *Der Thüringer Wald.*

71. ThStAGotha, Geh. Archiv, UU I,1, 31–35. These documents, which include notice of Seckendorff's official appointment as *Geheimer Rat*, indicate that he was head of the Gotha Kammer from 1656 to 1663.

72. Accounts of Seckendorff and Gotha have relied overmuch on printed material, thereby missing the extent to which Duke Ernst expected Seckendorff to produce good publicity for his fledgling duchy. See Ruge, "Vom Bibliothekar zum Geheimen Rat," 2–20.

73. That Seckendorff was miserable in his duties as director of the Kammer and secret councillor is a very recent discovery, for which we have Roswitha Jacobsen to thank. Seckendorff wrote about his troubles only after retiring to his private estate, Meuselwitz, in Zeitz. The document is near there, in Altenburg, but I found it difficult to decipher. Happily, it has been transcribed. See Jacobsen, "Die Brüder Seckendorff," 117–19, 117; and ThStAAltenburg, Nr. 1066, 1–6.

74. This, at least, was clear without the records from Seckendorff's family archive. See reports from his trips with the princes. ThStAGotha, E IV §, Nr. 2a and E IV §, Nr. 5.

75. The records that detailed daily administration in the Kammer and the secret council were destroyed in 1855. A marginal note in a "Findbuch" in Gotha reads: "They [the minutes of the Kammer and secret council] were in the archive vault in 1848, on the right near the entrance, and were destroyed on the order of State Minister his Excellence von Seebach at the beginning of 1855. Beck." It is beyond strange that these absolutely vital documents, from the very heart of state administration during the first years of the duchy's existence, were thrown away. (Hans-Jörg Ruge, a scholar-librarian in Gotha, confirmed this when we spoke about it on 24 June 2004.) The "Beck" who signed off on that order was August Beck, the same man who established Sachsen-Gotha as the model of the well-ordered Lutheran principality and Duke Ernst as its pious architect. His books started appearing during the 1860s, not long after the archival documents in question were "*cassirt*." Only a few of the minutes from the Kammer in Friedenstein between 1683 and 1684 seem to have survived. See Beck, *Ernst der Fromme* (1865), *Geschichte der Stadt Gotha* (1870), and *Geschichte des gothaischen Landes* (1860). Beck apparently did not bother to check the Seckendorff family archive in Altenburg for offending documents (ThStAAltenburg, Nr. 1066).

76. Small, *Cameralists*, 60–69.

77. Ibid., 285.

78. Ibid., 285

79. See, for example, Tribe, *Governing Economy*, 275; Adam, *J. H. G. Justi*, 11–12; Klein, "Johann Heinrich Gottlob von Justi," 145–47.

80. Vierhaus, *Germany in the Age of Absolutism*, vii.

81. On the Reich as "incubator," see Walker, *Home Towns*, 11–33. See also Aretin, *Heiliges Römisches Reich*, vol. 1; and Gagliardo, *Reich and Nation*, 9–10.

82. For the state-of-the-art textbook account, see Sheehan, *German History*, 193–95.

83. On cameralism as "descriptive," see Friedrich, "Continental Tradition," 130; and Johnson, "Concept of Bureaucracy," 379.

84. Schumpeter, "The Crisis of the Tax State," 19.

85. See, for example, Ferguson, *The Cash Nexus*, 54–55; and Bonney, ed., *Rise of the Fiscal State*.

86. For an alternative view, see Ogilvie, "The State in Germany: A Non-Prussian View."

87. Scott, *Seeing Like a State*, 11–22. Scott's narrative leans heavily on Lowood, "The Calculating Forester."

88. Research from the past several decades has already undermined many traditional clichés about the nature of German, and especially Prussian, state formation. See, for example, Neugebauer, *Politischer Wandel im Osten;* and Ogilvie, "Germany and the Seventeenth-Century Crisis."

89. This is exhaustively detailed in Ashworth, *Customs and Excise*.

90. See Blackbourn and Eley, *Peculiarities of German History*, 10–11.

91. Schmoller, ed., *Staatsverwaltung Preussens;* also Hintze, *Regierung und Verwaltung*.

92. Some recent contributions in this vein include Ferguson, *The Cash Nexus;* MacDonald and Gastmann, eds., *History of Credit and Power;* and Greenfeld, *Spirit of Capitalism*.

93. Schumpeter, "Crisis of the Tax State," 17.

94. See Frängsmyr, Heilbron, and Rider, eds., *Quantifying Spirit* Wise, ed., *The Values of Precision;* Porter, *Trust in Numbers;* Hessenbruch, "The Spread of Precision Measurement."

95. See Ziolkowski, *German Romanticism and its Institutions*.

96. Jacob and Stewart, *Practical Matter*, 157; Mokyr, "Intellectual Origins."

97. Mokyr, "Intellectual Origins," 336.

CHAPTER TWO

1. Schemnitz, known today as Banská Štiavnica (Slovakia), formed the administrative heart of an important Habsburg mining district.

2. See Rudwick, *Limits of Time,* 22–27.

3. See Bartels, *Montangewerbe zur Bergbauindustrie,* 334–53.

4. Seckendorff, *Teutscher Fürsten Stat,* 1:405–6.

5. Justi, *Abhandlung von den Mitteln die Erkenntniß,* 12.

6. Small, *Cameralists,* 285; Wagenbreth, *Die technische Universität,* 27; Smith, *Business of Alchemy;* Roscher, *Nationalökonomik,* 1:294; Hörnigk, *Oesterreich über alles.*

7. See Frensdorff, "Das Leben," 440–59. Justi's cameralist contemporaries, Daniel Gottfried Schreber and Georg Heinrich Zincke, also had connections to mining and metallurgy. See Schreber, *Neue Sammlung;* Zincke, *Cameralisten-Bibliothek.*

8. See Cohen and Wakefield, eds., "Introduction," in Leibniz, *Protogaea;* Hamm, "Knowledge from Underground," 77–99; Cohen, "Leibniz's *Protogaea,*" 42–59; Steenbuck, *Silber und Kupfer aus Ilmenau;* Wagenbreth, *Goethe und der Ilmenauer Bergbau;* Jackson, "Natural and Artificial Budgets"; Schellhas, "Abraham Gottlob Werner als Inspektor"; Ospovat, "The Importance of Regional Geology," and "Romanticism and Geology"; Laudan, *From Mineralogy to Geology,* 87–112; and Baumgärtel, "Alexander von Humboldt."

9. Bozorgnia, *Precious Metals,* 164–65. Some of the Austrian Habsburg possessions that lay outside the borders of the empire were rich sources of silver and other minerals, but these too were administered by mining officials who reported to Vienna.

10. See Soetbeer, *Edelmetall-Produktion;* and Nef, "Silver Production in Central Europe, 1450–1618," 575–91. Although Sweden's silver production remained well below that of Saxony, Hannover, or Austria, Sweden's mining administration provided an important model for central Europe's cameralists, especially with regard to iron production. See Hessenbruch, "The Spread of Precision Measurement."

11. Bozorgnia, *Precious Metals,* 165.

12. Soetbeer, *Edelmetall-Produktion,* 70–71; Bozorgnia, *Precious Metals,* 168–70.

13. Bozorgnia, *Precious Metals,* 169; Nef, "Mining and Metallurgy in Medieval Civilization."

14. Baumgärtel, "Absolutismus," 26–33.

15. Bartels, *Erzbergbau im Oberharz,* 46–86.

16. Baumgärtel, "Absolutismus," 10. During the last half of the eighteenth century, mining officials moved to exploit other sources of mineral wealth, especially in Prussia. Even so, contemporaries recognized that the silver mines continued to hold a privileged position. See Justi, *Manufacturen und Fabriken,* 2:253.

17. In the Harz Mountains, where it became common for mining officials to own shares in the mines they administered, foreign *Gewerken* (i.e., groups of investors) played a smaller role than in Saxony. See Bartels, *Erzbergbau im Oberharz,* 476; and Kraschewski, "Zur Arbeitsverfassung des Goslarer Bergbaus," 275–304.

18. There were also significant technological and economic reasons for the con-

centration of mining operations. See Wagenbreth and Wächtler, eds., *Der Freiberger Bergbau.*

19. On Saxony, see Köhler, *Bergmännischer Kalendar;* Baumgärtel, "Absolutismus," 25–51; and Weber, *Heynitz,* 132–33, 137–41. On the Harz, see Bartels, *Erzbergbau im Oberharz;* and Kraschewski, "Das Direktionsprinzip im Harzrevier."

20. In Saxony the most important of these levies was the *Zehnt,* which, according to tradition, bound the *Gewerken* to pledge a tenth of all ore as tribute to the sovereign. Though the *Zehnt* hovered around 10 percent of total silver yield for centuries, it generally amounted to less than that since one had to make concessions to the *Gewerken* for maintaining the "contribution mines" (*Zubußgruben*), that is, for maintaining unprofitable mines and infrastructure. See Köhler, *Bergmännischer Kalendar,* 77–97; Baumgärtel, "Absolutismus," 31–33.

21. Johann Köhler argued that the *Direktionsprinzip* grew naturally out of the *Bergregal,* or mineral privilege of the sovereign. Baumgärtel, for his part, holds that the *Direktionsprinzip* arose out of the special financial demands and administrative circumstances of the mines. See Köhler, "Die Keime des Kapitalismus," 106–7; Baumgärtel, "Absolutismus," 25–26.

22. Fritzsch and Sieber, *Trachten,* 14.

23. Heynitz to Elector Friedrich August, 8 August 1774. StADresden Loc. 1327, 353.

24. According to Trebra, the mining parades became increasingly regimented and militaristic under Heynitz. See Trebra, *Bergmeister Leben,* 146–56.

25. Heynitz's original *Bergbarte* is displayed at the Bergbau Museum in Freiberg.

26. Köhler, *Bergmännischer Kalender,* 1:28–29.

27. Ibid., 1:34.

28. Cobalt smuggling, for example, became so bad that Saxony condemned smugglers to death in 1723; see Lünig, ed., *Codex Augusteus,* 2:377–488.

29. I have here adopted the language of the Kammer and its officials. But there is another way to look at it: perhaps mining officials, with their new orders and regulations, were intruding on the customary practices of an already existing "common economy." Cf. Ashworth, *Customs and Excise,* 154–57.

30. Seckendorff, *Teutscher Fürsten Stat,* 1:394–416; 1:405–6; 1:411.

31. It remains a matter of debate how much impact such mining officials actually had. See Weber, *Heynitz,* 128.

32. After Heynitz resigned from Saxon service in 1775, he left for Paris, where he served as an advisor to the mint and wrote on political economy. He published the *Essai d'économie politique* anonymously in 1785. It was translated into German the next year, where it appeared, also anonymously, as *Tabellen über die Staatswirthschaft eines europäischen Staates der vierten Größe.*

33. Heynitz, *Tabellen,* preface.

34. Justi, ed., *Policey-Amts Nachrichten* (1756), 93.

35. Ibid., 89.

36. Ibid., 89.

37. See, for example, the contributions in Wächtler and Engewald, eds., *Internationales Symposium.*

38. Cited in Schlechte, *Staatsreform,* 24.

39. [Justi], *Brühl;* Böttiger, *Sachsen,* 2:438–53.

40. These memoranda are collected in Schlechte, *Staatsreform.*

41. Brühl to Fritsch, 15 March 1762, in Schlechte, *Staatsreform,* 270.

42. Memorandum from Fritsch to Brühl, 29 March 1762, in Schlechte, *Staatsreform,* 218–20.

43. On Heynitz's appointment and activities in Saxon Service, see Weber, *Heynitz,* 116–67; Baumgärtel, "Absolutismus," 67–99; and Schlechte, *Staatsreform,* 72–75.

44. Weber, *Heynitz,* 116–19.

45. StADresden, Loc. 1327, 1–7.

46. Weber, *Heynitz,* 120–21.

47. StADresden, Loc. 1327, 21–24. See also Weber, *Heynitz,* 157.

48. StADresden, Loc. 1327, 21–22.

49. StADresden, Loc. 1327, 22.

50. Ohain came from an old family of elite Saxon mining officials. Cf. Gleeson, *Arcanum,* 10–11.

51. StADresden, Loc. 1327, 22–23.

52. StADresden, Loc. 1327, 22–23; my emphasis.

53. StADresden, Loc. 1327, 24.

54. Heynitz to Elector Friedrich August, 27 January 1769. StADresden, Loc. 36216, 2.

55. Weber, *Heynitz,* 156–57.

56. Both would later become teachers at the mining academy.

57. StADresden, Loc. 514, 1–6.

58. See Baumgärtel, "Vom Bergbüchlein zur Bergakademie," 142–45; Weber, *Heynitz,* 156.

59. Heynitz's reference here echoes the concerns of officials in Göttingen, Lautern, and Ingolstadt about useless projectors (see chapters 3, 5, and appendix 3.) On the linkages between alchemy, fraud, medicine, speculation, and useful knowledge, see Gleeson, *Arcanum,* 79–84; Smith, *Business of Alchemy,* and Cook, *Matters of Exchange,* 379–409.

60. See Heynitz to elector Friedrich August, 12 April 1771. StADresden, Loc. 1327, 121–33; Weber, *Heynitz,* 157.

61. Though many of the Bergakademie's students also attended university, the academy began to usurp some of the functions that university education had once provided.

62. See Meinel, "Reine und angewandte Chemie," 25–45.

63. StADresden, Loc. 36216, 5.

64. Fritsch to Prime Minister Brühl, November, 1761, in Schlechte, *Staatsreform,* 181.

65. Heynitz was very much concerned with details of the mining uniforms. See Trebra, *Bergmeister Leben,* 125–29.

66. StADresden, Loc. 514, 61; UniArchFreiberg, Sekt. 91d, Acta 8045; Kern, *Bericht vom Bergbau* ("Avertissement").

67. Some have noticed this correlation (Weber, *Heynitz,* 152–54), but most (Cf. Baumgärtel, "Bergbüchlein zur Bergakademie") have tended to overlook it. There is a substantial body of scholarship on the establishment and early history of the Bergakademie Freiberg; see, for example, Papperitz, *Bergakademie zu Freiberg;* Herrmann, *Bergbau und Bergleute;* Hoffmann, *Bergakademie Freiberg.* On Schemnitz, see Faller, *Berg- und Forst-Akademie in Schemnitz.* On the Bergakademie in Berlin, see Strunz, *Von der Bergakademie zur Technischen Universität Berlin,* 11–17. On the mining school in Clausthal, see *Festschrift zur 175-Jahrfeier der Bergakademie Clausthal.*

68. Anon., *Der Nutzen und die Nutzungen des Bergwerks;* Henckel, *Kleine minerologische und chymische Schrifften,* preface.

69. Henckel, *Kleine minerologische und chymische Schrifften,* preface; and Zimmermann, *Obersächsische Berg-Academie.* Henckel and Zimmermann worked closely together, and it is possible that many of the plans laid out in *Obersächsische Berg-Academie* originated with Henckel, himself an early proponent of mining academies. See Herrmann, *Bergrat Henckel.*

70. Zimmerman, *Obersächsische Berg-Academie,* 5–6, 129–31. Baumgärtel has shown the significance of Zimmermann's *Obersächsische Berg-Academie* for Heynitz, who was studying privately in Freiberg with Mining Councilor Henckel during the early 1740s (see Baumgärtel, *Bergbüchlein zur Bergakademie*).

71. Justi, *Grundsätze der Policey-Wissenschaft,* 98; Weber, *Heynitz,* 154.

72. Schreber, *Sammlung verschiedener Schriften* (1763): 10:417–36 ("Entwurf von einer zum Nutzen eines Staats zu errichtenden Academie der öconomischen Wissenschaften"); Stieda, *Nationalökonomie als Universitätswissenschaft,* 55; Weber, *Heynitz,* 154; Tribe, *Governing Economy,* 91–94.

73. As in Freiberg, these mining academies did not arrive suddenly. In Schemnitz, for example, the mining sciences had been taught for many years to apprentices. The move to an academy there had many of the same implications that it did in Freiberg. Austria's mining administration was moving to train a new generation of cameralist mining officials. See Faller, *Berg- und Forst-Akademie in Schemnitz,* 4–13. In Berlin the case was similar, though Prussia made more of an effort than most other territories to train its officials in the universities.

74. GStAPK, I HA Rep. 121, Abt. D, Tit. II, Sect. 1, Nr. 101, Bd. 1, 33. Gerhard submitted a detailed report about Freiberg to Minister Hagen in the spring of 1770. "Since Your Excellency has ordered me to gather information on the arrangements of the mining academy established in Freiberg in 1766, so I have humbly to report that at

said academy there are 2 professors, namely Mining Councilor Gellert and Professor Charpentier."

75. See, for example, ArchBreslau, Sygn. 00303, Nr. 406; GStAPK, I HA Rep. 121, Abt. D, Tit. II, Sect. 1, Nr. 101, Bd. 1, 33.

76. On Justi's disastrous tenure as Berghauptmann in Prussia, see chapter 4.

77. The order from Berlin appeared in the form of a 12 December 1768 edict specifying that the following subjects should be taught at Frederick's universities: "jus metallicum, historiam naturalem, mineralogiam, metallurgiam, chemiam, mathesin puram et applicatam, physicam theoreticam et experimentalem." A later rescript from the General Directory to minister von Fürst added "mining jurisprudence" to the list of required subjects. GStAPK, I HA Rep. 121, Abt. D, Tit. II, Sect. 1, Nr. 101, Bd. 1 , 10.

78. Ibid.

79. Ibid.

80. Delius, *Anleitung*, preface.

81. Delius, "Abhandlung," 3; in *Anleitung*. (The special treatise, or "Abhandlung," on mining-cameral-science that appeared at the end of the textbook had its own pagination.) Cf. Rudwick, *Limits of Time*, 24–26.

82. Delius, *Anleitung*, dedication.

83. Ibid., preface.

84. Hannover's first mining academy was established in Clausthal in 1853, though there had been a mining school there since 1775. See *Festschrift zur 175-Jahrfeier der Bergakademie Clausthal.*

85. On Münchhausen and Göttingen, see chapter 3.

86. See chapter 3.

87. See Kästner, *Anmerkungen über die Markscheidekunst;* and Gmelin, *Chemische Grundsätze der Probir- und Schmelzkunst.*

88. In the original, the entry reads: "Von dem Churfürst. Sächs. Oberbergdirektor von Heinitz geschenkt." Note his designation as "Oberbergdirektor" instead of the proper "Oberbergkommissar." See HandAbtUniGö, Bibl. Arch., Manuale 1769, p. 93, accession no. 4.M.8788.

89. On Heyne's approach to book acquisition and its reflection of Göttingen's larger interests, see his reminiscences in the *Göttingische gelehrte Anzeigen.* These books appear in Otfried Wagenbreth's compilation of the "most important literature" of the early modern mining sciences (Wagenbreth, *Technische Universität,* 27). On the history of the Göttingen collections, see Reimer Eck, "Göttingen University Library: Excursions into the Archives."

90. The 2 March 1771 final report, or *Revisionsbericht,* of the Audit Commission is housed in StADresden, Loc. 1327, 135–96. It is reprinted in Baumgärtel, "Absolutismus," 127–92. The page numbers in subsequent citations refer to Baumgärtel's published reprint.

91. In fact, however, as Weber has demonstrated, Heynitz remained an outsider

in Dresden's *Berggemach,* which ultimately hampered his ability to implement the changes he desired.

92. *Revisionsbericht,* 2 March 1771, 127–28. Hereafter in this chapter cited by page number in text.

93. Trebra, *Bergmeister Leben,* 20–22. Hereafter in this chapter cited by page number in text.

94. UniArchFreiberg, Akte OBA 7917, 1:231–33.

CHAPTER THREE

1. [Böll], *Das Universitätswesen in Briefen,* 4. For another analysis of this passage, see Clark, *Academic Charisma,* 379–80. In Clark's formulation, Böll's comparison is "hopefully a satire." Satire or not, the Kammer in Hannover took this comparison quite literally.

2. [Böll], "Bemerkungen über Johann Jacob Mosers Rede, wie Universitäten, besonders in der juristischen Facultät, in Aufnahme zu bringen und darinn zu enthalten," in Rössler, *Gründung,* 468–86. Martin Gierl identified Böll as the probable author of the "Bemerkungen" (see Gierl, "Aufklärungsfabrik"). The "Bemerkungen" became so strongly identified with Münchhausen that his biographer, Walter Buff, put Böll's words in the curator's mouth, even though he recognized that the "Bemerkungen" were not actually written by Münchhausen (Buff, *Münchhausen,* vi, 38, 56, 81, 126, and 133). Later works simply attributed the "Bemerkungen" to Münchhausen. See Schelsky, *Einsamkeit und Freiheit,* 31–32. On Böll, Göttingen, and "academic cameralism," see Clark, *Academic Charisma,* 373–81.

3. Michaelis, *Raisonnement,* 1:1–5.

4. There is an extensive literature on the University of Göttingen. Many contemporaries, especially professors, wrote about the university. See Pütter, *Versuch einer academischen Gelehrten-Geschichte;* Brandes, *Ueber den gegenwärtigen Zustand der Universität Göttingen;* Meiners, *Geschichte der Entstehung und Entwicklung der hohen Schulen;* and Michaelis, *Raisonnement.* Later accounts tended to view Göttingen in the light of the University of Berlin. See Rössler, *Gründung;* Paulsen, *Das deutsche Bildungswesen,* and *Geschichte des Gelehrten Unterrichts;* Selle, *Die Georg-August-Universität.* For more recent perspectives, see Turner, "The Prussian Universities and the Research Imperative"; McClelland, *State, Society, and University;* and Clark, *Academic Charisma,* 247, 377–81. Johann Christoph Erich Springer's anonymous and sarcastic critique of Göttingen, *Über die protestantischen Universitäten in Deutschland,* referred to Göttingen as a "fiscal university." Turner called state policy toward the university "academic mercantilism" in his dissertation, "Prussian Universities and the Research Imperative," 23, 68–70.

5. Pütter, *Versuch einer academischen Gelehrten-Geschichte,* 2:13.

6. Münchhausen was appointed president of the Kammer in 1753. King George II called him "mon président des finances." See Bärens, "Kurtze Nachricht," 66, n.

3. On Münchhausen, see Rössler, *Gründung;* Buff, *Münchhausen*; Clark, *Academic Charisma,* 246, 380; Black, "Hanoverian Nexus."

7. Much of the literature touching on the Münchhausen brothers, especially English-language scholarship, has focused on diplomatic and political history. See Dann, *Hanover and Great Britain;* and Riotte, "George III and Hanover."

8. StArchHan Dep. 7 B, Nr. 2043, 50–57, 64.

9. There was concern from the very beginning about adequate housing, reasonable prices, and suitable amenities. Most of this fell under the domain of "good police." Münchhausen, especially, dedicated himself to this effort. Rössler, *Gründung*, 66–74, 385–91, 415.

10. Münchhausen, quoted in Selle, *Die Georg-August Universität*, 49–53. On the University of Göttingen's relationship to England and English students, see Biskup, "University of Göttingen."

11. These officially-sanctioned advertisements included the anonymous, three-volume *Zeit- und Geschicht-Beschreibung der Stadt Göttingen* and Claproth's 1748 work, *Der gegenwärtige Zustand der Göttingischen Universität.* Münchhausen personally directed Claproth, and then Justi, to publicize the wonders of Göttingen; see HandAbtUniGö, 2 Cod. MS Michael. 324. See also Sachse, *Göttingen im* 18. *und* 19. *Jahrhundert*, 57–58.

12. Anon., *Zeit- und Geschicht-Beschreibung,* vol. 1, preface.

13. See [Claproth], *Der gegenwärtige Zustand der Göttingischen Universität*, 8–9.

14. "Es haben mir Herrn Cammer President Excellenz Auftrag gethan, ein Schreiben von dem Zustand der Universität, wie ehedem der Sec. Claproth, drücken zu laßen, und darüber mit deroselben zu communiciren." Justi to Michaelis, 13 April 1756. HandAbtUniGö, 2 Cod. MS Michael. 324, fol. 490. Justi suggested to Michaelis that he would begin drafting the work during his upcoming vacation at the baths. I have been unable to locate any sign of it.

15. Bärens, "Kurtze Nachricht,"55–59.

16. Bärens, "Kurtze Nachricht," 58–60. See also Mohnhaupt, *Die Göttinger Ratsverfassung*, 108–17.

17. Camelott, a double-woven woolen fabric, was originally produced in parts of Belgium. Camelott was made with carded wool and a chain of worsted yarn. According to Prussian regulations, each piece of Camelott had to be thirty ells long and one ell wide. See Justi, *Manufacturen und Fabriken,* 2:50–51; and Krünitz, *Oeconomische-technologische Encyklopädie*, "Camelott."

18. By way of comparison, consider that annual maintenance costs for the entire university were 16,600 taler in 1733. See Rössler, *Gründung*, 57; Bärens, "Kurtze Nachricht," 64.

19. Koch, *Das Göttinger Honoratiorentum,* 119.

20. Ferdinand Frensdorff, who wrote much about Göttingen, found this assertion so jarring that he made a special point to reject it out of hand. See his introduction to Bärens, "Kurtze Nachricht," 53.

21. Koch guessed that the memo was written by a town councilor named Sothe, who was responsible for overseeing manufactures in Göttingen during 1724–25. Koch, *Das Göttinger Honoratiorentum,* 61, n. 7. The memo is still in the Göttingen City Archive today, but under a different call number than during Koch's time: StadtArchGö AA, Industrie, Fabrik- und Manufaktursachen, Nr. 15 (hereafter referred to as the "1724 Memorandum" and cited in the text by section number; e.g. "sec. 5").

22. Though Göttingen had begun to experience some economic recovery by 1720, the town still appeared stagnant and run-down to contemporary observers. (See Winnige, *Krise und Aufschwung*, 406.) Göttingen's construction boom did not really begin in earnest until the 1730s, during the first years of the university. See Kastner, "Wohnen und Bauen in Göttingen," 182–205; and Winnige, *Krise und Aufschwung,* 258.

23. See also Winnige, *Krise und Aufschwung,* 126–33.

24. Around 1750 Göttingen had 424 officially sanctioned brew houses and about 390 burghers with brewing rights among its roughly 8,600 residents (Winnige, *Krise und Aufschwung,* 103, 330–33; and Corran, *A History of Brewing*, 46–47).

25. This may have been an overestimation. Winnige estimates the number of cloth makers at thirty-three in 1689 and seventy-four in 1763 (*Krise und Aufschwung*, 128–29).

26. The most important division existed between the *Tuchmacher* and *Zeugmacher*. The two guilds were closely related in terms of products and markets, which could make conflicts between them especially bitter. *Zeug*, which used carded wool, tended to be cheaper and lighter than *Tuch*, a heavier and purer variety of woolen cloth. See Bohnsack, *Spinnen und Weben*, 174.

27. 1724 Memorandum, sec. 7; Koch, *Das Göttinger Honoratiorentum,* 67.

28. The influence of the royal government, with its orders for large amounts of cloth, began to concentrate production in the hands of entrepreneurs like Ebel, Lüdeck, and Grätzel. But even in these larger enterprises, much of the actual work—the shearing, carding, spinning, weaving—was still performed in the homes of the workers. For the seminal work on proto-industry, see Kriedte, Medick, and Schlumbohm, *Industrialisierung vor der Industrialisierung*. See also Medick, *Weben und Überleben,* introduction.

29. Gallenkamp, like Grätzel, was an outsider. By 1756, when Justi was in Göttingen, only nineteen independent guild masters remained, while Grätzel had thirty-one cloth makers working for him. See Brückner et al., "Vom Fremden zum Bürger," 119; Winnige, *Krise und Aufschwung*, 129; Koch, *Das Göttinger Honoratiorentum,* 68–70; 1724 Memorandum, secs. 10–11.

30. Kastner, "Wohnen und Bauen," 185–93; Losch, "Die hessischen Prinzen in Göttingen," 28.

31. This proposal had lasting appeal. Münchhausen himself, for example, would become involved in establishing a lending house for Göttingen's manufacturers, and

he attempted to found a "woolens magazine" where producers could store their wares. See Rössler, *Gründung*, 415.

32. In the decade after 1724 Hannover devoted special attention to this policy, working with various royal "factors" and "commissars" to ensure that the army got its cloth from Göttingen. This was the single largest reason for the growth of Göttingen's cloth manufactures after 1725.

33. Höttemann, "Die Göttinger Tuchindustrie," 46–47; Klammer, *Gewerbeentwicklung*, 50–51; and StadtArchGö, AA, Abgaben, Onera, Nr. 58.

34. Grätzel arrived in Göttingen in 1711 and initially worked as an apprentice for Ebel and Lüdeck. See StadtArchGö, AA, Abgaben, Onera, Nr. 58; Koch, *Das Göttinger Honoratiorentum*, 68, 100; Bärens, "Kurtze Nachricht," 64; Wagner, "Ober-Commissarius Graetzel," 79.

35. Dyeing with woad was no small feat. Decades later, for example, Johann von Justi claimed to have discovered the "secret" of dyeing with woad, and he published several treatises on the subject. See Justi, *Policey-Amts Nachrichten* 1 (1755): 1–27.

36. In 1723, when his manufactory was just starting up, Grätzel coaxed a substantial five hundred taler out of the Kammer. See Koch, *Das Göttinger Honoratiorentum*, 68–69.

37. Brückner et al., "Vom Fremden zum Bürger," 119; Koch, *Das Göttinger Honoratiorentum*, 68–69.

38. Grätzel's success, which depended on restructuring the entire system of production and delivery, harmed not only the cloth-makers' guild, but the merchants' guild as well. See Koch, *Das Göttinger Honoratiorentum,* 48–135.

39. Ibid., 71.

40. Ibid., 102–3.

41. Extract of 13 April 1732. See StadtArchGö, AA, Abgaben, Onera, Nr. 58; Koch, *Das Göttinger Honoratiorentum*, 105.

42. Manufacturers generally paid only 3 percent annual interest for these loans (Winnige, *Krise und Aufschwung*, 395–404). By way of comparison, the average value of a house in Göttingen was eight hundred taler in 1765. See Gerhard, "Geld und Geldwert im 18. Jahrhundert," 29.

43. Koch, *Das Göttinger Honoratiorentum*, 106.

44. Höttemann, "Göttinger Tuchindustrie," 60.

45. Rössler, *Gründung*, 415.

46. Koch, *Das Göttinger Honoratiorentum*, 106.

47. Michaelis, *Raisonnement,* 1:1–5.

48. George II appointed Münchhausen to the secret council (*Geheimer Rat*) in 1728. He was subsequently appointed as *Großvogt* in Celle (1732), *Kammerpräsident* in Hanover (1753), and prime minister (1765). See Dieter Brosius, "Gerlach Adolf Freiherr v. Münchhausen," 523–24.

49. Wagnitz, *Zuchthäuser in Deutschland,* 67.

50. HStArchHan, Dep. 7 B, Nr. 237, 12–14.

51. On this legend of Celle's "Zucht-, Werk-, und Tollhaus," see Harms, "Editorial."

52. [Anon.], *Zeit- und Geschicht-Beschreibung,* 2:115.

53. Kastner, "Wohnen und Bauen," 196–206.

54. Grätzel to Hannover, 21 January 1734 (StadtArchGö, AA, Abgaben, Onera, Nr. 58).

55. Hannover to mayors and town council of Göttingen, 24 February 1734 (ibid.).

56. A draft of the ordinance appeared along with the official correspondence from Hannover. See StadtArchGö, AA, Industrie, Fabrik- und Manufaktursachen, Nr. 17.

57. The report from Göttingen included many pages of samples, so that the Kammer could judge for itself which manufacturers merited support (ibid.).

58. Münchhausen to Göttingen town council, 2 May 1740 (ibid.).

59. Ibid.

60. Koch, *Das Göttinger Honoratiorentum,* 112.

61. Fuller's earth, a hydrous silicate of alumina, was used to cleanse cloth (ibid., 122).

62. Hannover to Göttingen town council, 17 August 1739 (StadtArchGö, AA, Abgaben, Onera, Nr. 58).

63. About Grätzel's house on the *Allee* (now "Goetheallee 6") see Kastner, "Wohnen und Bauen," 245–46; Koch, *Das Göttinger Honoratiorentum,* 115–16; Wagner, "Der Ober-Commissarius Graetzel," 80.

64. Minutes from a report by Göttingen's deputies to Hannover's secret council, 9 December 1743 (StadtArchGö, AA, Abgaben, Onera, Nr. 58).

65. Münchhausen, on behalf of the secret council, to Göttingen town council, 19 December 1743 (ibid.).

66. Münchhausen to Göttingen town council and mayors, 17 August 1747 (ibid.).

67. Münchhausen brought in a relative, Oberhauptmann Borries von Münchhausen, to help resolve the dispute (ibid.).

68. Seidler to Münchhausen, 17 December 1744 (ibid.).

69. Gesner to Grätzel, 27 September 1744 (ibid.).

70. Gesner to Münchhausen, 17 December 1744; Seidler to Münchhausen, 17 December 1744 (ibid.).

71. Grätzel to Münchhausen, 28 December 1744 (ibid.).

72. Grätzel to Privy Secretary Unger, 28 December 1744 (ibid.).

73. Grätzel to Münchhausen, 28 December 1744 (ibid.).

74. Hannover to Göttingen deputies (*Bürgerdeputierte*), 10 January 1745 (ibid.).

75. Göttingen's town and guild representatives to Hannover, 13 January 1745 (ibid.).

76. Grätzel had already been in Göttingen since 1711, much longer than twenty-two years.

77. Göttingen's town and guild representatives to secret council, 13 January 1745 (StadtArchGö, AA, Abgaben, Onera, Nr. 58).

78. Hannover to Göttingen town council, 26 January 1744 (ibid.)

79. Mayor Insinger, who had been installed by the royal government, was a friend of Grätzel's. It was "well known," complained one embittered merchant, that he was Grätzel's "intimate friend." See Mohnhaupt, *Göttinger Ratsverfassung*, 108–17; and Koch, *Das Göttinger Honoratiorentum*, 112.

80. There are those who would consider this a fair description of present-day Göttingen, but things really have improved.

81. Rössler, *Gründung*, 70.

82. Koch, *Das Göttinger Honoratiorentum*, 111.

83. Rössler, *Gründung*, 401; Wagner, "Der Ober-Commissarius Graetzel," 79.

84. Bärens, "Kurtze Nachricht," 73; Wagner, "Der Ober-Commissarius Graetzel," 79.

85. This may have been Johannes Roszfeld (1550–1626), who wrote about Roman antiquities.

86. Koch, *Das Göttinger Honoratiorentum*, 113.

87. The university eventually acquired Grätzel's cabinet and put it in the library museum. It remained an important part of the university's collections well into the nineteenth century. See Selle, *Georg-August-Universität,* 307–8.

88. "Landesherrliche Verordnung vom 19. August 1743 wegen Eröffnung eines Fabrik Gerichts in Göttingen." Copy in StadtArchGö, AA, Abgaben, Onera, Nr. 58.

89. Ibid.

90. Koch, *Das Göttinger Honoratiorentum*, 108–126; and Höttemann, "Göttinger Tuchindustrie," 61.

91. HStArchHann, Hann. 80 Hildesheim I, F, Nr. 109.

92. At least two Göttingen professors, Johann von Justi and Anton Friedrich Büsching, made special mention of Scharff and his manufactory. See Justi, *Manufacturen und Fabriken*, 2:30 and 47; and Büsching, *Erdbeschreibung*, 3:3355.

93. Bärens, "Kurtze Nachricht," 64–65; Koch, *Das Göttinger Honoratiorentum*, 127.

94. Royal patent issued by George II, 18 October 1754, Kensington, England. The copy in the *Hauptstaatsarchiv* Hannover, which was sent to the secret council from the king, is signed by George II and countersigned by Münchhausen. HStArchHann, Hann. 80 Hildesheim I, F, Nr. 109. Camelott was made with carded wool and a chain of worsted yarn. *Barracan* was a double Camelott. See Justi, *Manufacturen und Fabriken,* 2:50–51; Krünitz, *Oeconomische-technologische Encyklopädie*, "Camelott."

95. Princess Maria, married to Prince Friedrich of Hessen-Kassel, was the daughter of George II.

96. Losch, "Die hessischen Prinzen," 28–33.

97. Ibid., 32.

98. On the enmity between Scharff and Grätzel, see Bärens, "Kurtze Nachricht," 64–65; Losch, "Die hessischen Prinzen," 33.

99. Scharff's manufactory, which was doing very well in 1755, gradually declined after that. By 1773 it was dead. See Koch, *Das Göttinger Honoratiorentum*, 126.

100. Secret council to Scharff, 3 July 1755; Secret council to Grätzel 17 July 1755. During his own fight with Widow Scharff many years later, Grätzel's son, Johann Heinrich (junior), included copies of these letters in his correspondence with the secret council. Grätzel (junior) to the secret council, 18 November 1780. See HStArchHann, Hann. 80 Hildesheim I, F, Nr. 109.

101. Rescript issued by Münchhausen (secret council), 18 November 1755. Copy included as "Enclosure E" in letter from Grätzel (junior) to the secret council, 18 November 1780. See HStArchHann, Hann. 80 Hildesheim I, F, Nr. 109.

102. Justi, ed., *Neue Wahrheiten*, vol. 1 (1754). His early work on the cameral sciences appeared just around this time as well. See Justi, *Gutachten von dem vernünftigen Zusammenhange und practischen Vortrage aller Oeconomischen und Cameralwissenschaften.*

103. Justi's letter to Münchhausen has apparently been lost, but his article on the simplified smelting process appears in his *Gesammlete Chymische Schriften.* See also his *Abhandlung über die Frage: Wie die Kupferertze . . . besser bearbeitet werden können.* The information here is based on a letter dated 24 October 1754 from a Hanoverian mining official to "*Kammerpräsident* von Münchhausen." Some of Justi's published works were included with the report, but I could not determine which ones; see UniArchGö, Kur 5.c.8, 26–27.

104. The *Kommunion Harz* was a mining district administered jointly by various lines of the Guelph dynasty. Its centers, after 1735, were the mining towns of Clausthal (Hannover) and Zellerfeld (Braunschweig-Wolfenbüttel). See Kraschewski, "Das Direktionsprinzip im Harzrevier"; Henschke, *Landesherrschaft und Bergbauwirtschaft,* 75–77; and Bartels, *Montangewerbe zur Bergbauindustrie*, 152–69.

105. Thanks to Hans Hofmann, archivist at the Bergakademie in Freiberg, for his help on identifying assorted mining officials. Unfortunately, I have been unable to decipher the names of Münchhausen's two informants here. One letter was sent to Münchhausen from Regensburg on 21 October 1754 (UniArchGö, Kur 5.c.8, 34–36); the other from Clausthal on 24 October 1754 (UniArchGö, Kur 5.c.8, 26–27). Justi himself, in the dedication to the first edition of *Staatswirthschaft*, attributed the abrupt departure from Vienna to illness.

106. The preface to the work is dated 11 April, 1755. Münchhausen sent the official offer to Justi on 29 April, 1755 (UniArchGö, Kur 5.c.8, 24).

107. Ibid., 24–25.

108. Justi to Münchhausen, 3 November, 1755 (UniArchGö, Kur 5.c.8, 21–22).

109. The Göttingen library retains a copy of Justi's invitation to his lectures.

110. Justi *Staatswirthschaft*, 1:x–xi.

111. Ibid., 1:xxii–xxiii.

112. Justi, *Abhandlung von den Mitteln die Erkenntniß in den Oeconomischen und Cameral-Wissenschaften,* 8.

113. Many eighteenth-century cameralists laid out plans for separate faculties and academies, especially after 1760. But Justi was an early proponent of institutionalization. The cameral academy in Lautern, which opened in 1774, was based on Daniel Gottfried Schreber's sketch from 1763. See Tribe, *Governing Economy,* 91–118; and chapter 5.

114. Justi, *Staatswirthschaft,* 1:xxxi–xxxiv.

115. UniArchGö, Kur 5.c.8, 5–6, 11.

116. [Claproth], *Schreiben von dem gegenwärtigen Zustande der Göttingischen Universität,* and UniArchGö, Kur 1.38.

117. Today's archival holdings, which scatter police records between the city and university archives, replicate the liminal status of police in 1750s Göttingen.

118. Justi's *Grundsätze der Policey-Wissenschaft,* his self-proclaimed "first system of police," established him as the foremost exponent of the police and cameral sciences. Considering that he wrote and published the work in Göttingen, very little has been written about his activities as chief police commissioner there. Even among those who devote attention to Justi's biography, the Göttingen period is largely ignored. Most convey the impression that Justi arrived in 1755, gave some lectures, and left two years later. See Adam, *J. H. G. Justi,* 39–41; Tribe, *Governing Economy,* 58; Lindenfeld, *Practical Imagination,* 25; Walker, *Home Towns*, 160. Even Ferdinand Frensdorff, whose biographical work on Justi remains the most thorough to this day, claimed that there was not much to say about Justi's activities as a police official in Göttingen. Frensdorff, "Die Vertretung der ökonomischen Wissenschaften," 520–22.

119. UniArchGö, Kur 5.c.10, 1–4.

120. I was unable to locate a copy of Justi's "Instruction," the document that delineated his powers and duties as *Ober-Policey-Commissar.*

121. UniArchGö, Kur 5.c.10, 7–9.

122. Ibid. The Kammer denied Justi's claim. "When it is stated in Section 23 of the Instruction: '*the vagabonds are to be arrested,*'" it explained, "such is to be understood *positis terminis habitibus,* that is, through the normal authorities." See UniArchGö, Kur 5.c.10, 5–6 (emphasis in original).

123. UniArchGö, Kur 5.c.9, 7–9.

124. Claproth to Justi (enclosed in Justi's complaint to the ministry), 19 November 1755 (ibid.).

125. Claproth to Hannover, 20 November, 1755 (ibid., 10–13).

126. Hannover to Justi, 24 November, 1755 (ibid., 1–3).

127. Justi to Münchhausen, 16 December, 1755 (UniArchGö, Kur 5.c.12).

128. Ibid., 10–11.

129. Frensdorff, "Vertretung der ökonomischen Wissenschaften," 520.

130. See chapter 2.

131. Justi, ed., *Policey-Amts Nachrichten*, (1755): 5–7, 73–75. See also Wakefield, "Police Chemistry."

132. UniArchGö, Kur 5.c.11, 1–4.

133. AkArchGö, Etat 6, 2, Nr.1–3.

134. The last entry in Justi's dossier, from 5 July 1757, grants him permission to travel to "Copenhagen for three months because of your health." He never returned (UniArchGö, Kur 5.c.8, 42).

135. I use the English term *manufactures* as a blanket term for what Justi called "*Manufacturen und Fabriken*." "*Manufacturen* and *Fabriken*," Justi confessed, "are usually taken to mean the same thing, and are used synonymously." He, however, chose to make a clear distinction between the two terms: *Fabriken* used "fire and hammer," whereas manufactures did not. See Justi, *Manufacturen und Fabriken*, 1:4–6.

136. The dedication to the first volume is dated 4 October 1757. Justi, *Manufacturen und Fabriken*, vol. 1, dedication.

137. Daß to P. F. von Suhm, 27 August 1757; quoted in Frensdorff, "Das Leben," 415.

138. There were many rumors about Justi's abrupt departure from Göttingen, among them debts, marital troubles, and danger of arrest. The immediate cause of his departure, however, was the advance of French troops into Göttingen. Due to Justi's past in Austria, he risked arrest by the French, who were allied with Austria. See Frensdorff, "Das Leben," 414–15.

139. Justi did indeed claim, in a work that appeared just before his trip to Denmark, that a "man should have more than one wife." See Justi, *Rechtliche Abhandlung von den Ehen*, 19. Isabell Hull has examined Justi's writings on marriage and sexuality. See Hull, *Sexuality, State, and Civil Society*, 179–97.

140. UniArchGö, Kur 4.V.b.35, 54–56. For the Swedish tradition linking cameralism and Linnaean botany, see Koerner, *Linnaeus*.

141. UniArchGö, Kur 4.V.b.35, 13–16, 39–44.

142. Hannover ignored Michaelis's request, and added his report to Springer's dossier.

143. Michaelis is referring to Daniel Gottfried Schreber, professor of œconomy in Bützow (UniArchGö, Kur 4.V.b.35, 57–59).

144. Ibid., 66–68.

145. Ibid., 60–65.

146. Schlözer went to Sweden in 1755, spent the winter of 1756–1757 in Uppsala, the following one and a half years in Stockholm, and then returned to Germany. See Pütter, *Versuch einer academischen Gelehrten-Geschichte*, 2:166.

147. Murray to Münchhausen, 5 April, 1767 (UniArchGö, Kur 4.V.b.36, 8–9).

148. Ibid.

149. Richter to Münchhausen, 6 April 1767 (ibid., 13–14).

150. Beckmann to Münchhausen, 16 April 1767 (ibid., 15–16).

151. Hannover to Beckmann, 19 April 1767 (ibid., 17–18).

152. Ibid., 5–6.

153. Hannover to Beckmann ("Pro mandatum Regis *p*"), 17 May 1770 (ibid., 25).

154. Beckmann, *Anleitung zur Technologie*, preface.

155. Beckmann, *Grundsätze der teutschen Landwirthschaft* (1769), and *Anleitung zur Technologie* (1777).

156. UniArchGö Kur 4.V.b.36, 43–44.

157. Ibid., 49–53. The position of *Hofrat* was regularly bestowed as a special honor on senior professors. In this case, Beckmann had received a special offer from Marburg, and Hannover felt the need to make a counter offer (58–60).

158. Justi to Michaelis, 13 April 1756. HandAbtUniGö, 2 Cod. MS Michael. 324, fol. 490.

CHAPTER FOUR

1. There is some debate about whether Justi actually died in prison (see Adam, *J. H. G. Justi,* 48), but it still seems most likely (see Beckmann, *Vorrath kleiner Anmerkungen,* 3:562; Frensdorff, "Das Leben," 457–59; Klein, "Johann Heinrich Gottlob Justi," 145; and Hubatsch, *Frederick the Great,* 84).

2. Frensdorff, "Das Leben," 441–42.

3. Hubatsch, *Frederick the Great,* 11–14.

4. Cf. Mittenzwei, *Preussen,* 223–30; Klein, "Johann Heinrich Gottlob Justi," 145–202.

5. Cf. Adam, *Justi,* 46–48; Remer, *Johann Heinrich Gottlob Justi,* 8–12. Remer looked (in vain) for exculpatory evidence in the archive and then claimed to have found it. Klein, in turn, used Remer's "evidence" to argue for Justi's innocence; see Klein, "Johann Heinrich Gottlob Justi," 145.

6. *Acta Borussica,* 14:443. Frensdorff's life of Justi is filled with asides about lazy biographers, like Johann Beckmann and a certain Madame D. M., who accepted rumor and second-hand accounts without checking them. Cf. Frensdorff, "Das Leben," 356–58, 439. Though Justus Remer went hunting for archival documents in the 1930s, he did not find much beyond what Frensdorff had already discovered (see Remer, *Johann Heinrich Gottlob Justi,* 11–12).

7. *Geheimes Staatsarchiv Preussischer Kulturbesitz, Berlin.*

8. Frensdorff, "Das Leben," 441–59.

9. These documents are housed in the Brandenburgisches Landeshauptarchiv Potsdam (now located outside of Potsdam, in Bornim), abbreviated here as LHAPotsdam.

10. Frensdorff, "Das Leben," 441.

11. Prussia was not known for its mines or its iron. Other regions of the empire, like the Harz and Erz mountains and Thüringen, had much stronger traditions. See Altmann, *Erzgebirgisches Eisen,* 167–77; and Ågren, "Swedish and Russian Iron-Making," 3–12.

12. See Justi, *Finanzwesens,* 128–31.

13. On the Prussian iron industry, See Weber, *Heynitz,* 168–236; and Beck, *Geschichte des Eisens,*3:334–45.

14. Wakefield, "Police Chemistry," 253–58.

15. Pott, *Chymische Untersuchungen,* 15–16.

16. Justi, *Chymische Schriften,* vol. 1, preface.

17. Ibid., 1:68–84, 87–94, 107–25; 2: 397–416.

18. See Wakefield, "Police Chemistry."

19. Justi, "Erweis, daß das Eisen nicht in dem Eisenerze, oder Steine, vorhanden sey."

20. Shapin and Schaffer, *Leviathan and the Air Pump,* 55–69.

21. Beck, *Geschichte des Eisens,* 3:69–71.

22. See Justi, *Chymische Schriften,* 1:79–83, 87–94, 95–106; 2:401–3.

23. Frederick II, *Anti-Machiavel.*

24. Though he tried to pass it off as an oversight or a slip, Justi's stance on tax farming and the excise in the *Finanzwesen* was completely consistent with his earlier work. Cf. his *Steuern und Abgaben,* 102–7; see also Adam, *J. H. G. Justi,* 222–31.

25. Justi, *Finanzwesens,* ix–xii.

26. Johnson, *Frederick the Great and His Officials,* 196–207; Schultze, *Regieverwaltung,* 31–71; *Acta Borussica,* 13:732, and 14:76–92, 177–87.

27. Justi, *Finanzwesens,* 58.

28. Ibid., 73–90. On the trope of "Plusmachen" in Justi's writings, see Adam, *J. H. G. Justi,* 188–94.

29. On projectors and *Plusmacher,* see Justi's *Gesammelte Schriften* 1:256–81 ("Gedanken von Projecten und Projectenmacher").

30. On Justi's cosmopolitanism, see Adam, *J. H. G. Justi.*

31. LHAPotsdam, Rep. 3, Nr. 4982, 135–39.

32. Ibid.

33. Ibid.

34. Justi was wrong about that. Even the most productive blast furnaces of the Ural region did not generally operate for a full forty weeks. Cf. Florén and Rydén, "Mines, Furnaces, and Forges," 84–87.

35. LHAPotsdam, Rep. 3, Nr. 4982, 139–40.

36. This is the title of his preliminary essay in the 1766 edition; see Justi, *Finanzwesen,* 28–44.

37. Ibid., 28–30.

38. Ibid., 30–31.

39. This was a common trope for Justi. Cf. *Staatswirthschaft,* preface; see also Frensdorff, "Das Leben," 460–62.

40. Justi, *Finanzwesens,* 34–35.

41. Justi, *Finanzwesens,* 37–38, 38–39.

42. Ibid., 39–42.

43. Ibid., 42–43.

44. LHAPotsdam, Rep. 3, Nr. 4884, 3. 2 October 1765, sent from Loben [Polish: Lubliniec], near Tarnowitz, in upper Silesia. Justi is already acting in his capacity as Berghauptmann.

45. During operation, ironmasters tried to keep the furnace in continual operation, that is, to avoid having it shut down, or "blow out," because restarting, or "blowing-in" a blast furnace demanded much time and fuel to expel moisture and prepare the furnace for smelting.

46. LHAPotsdam, Rep. 3, Nr. 4884, 43–45.

47. Ibid., 1–4.

48. The unit of measurement was a *Klafter,* a stack of wood roughly comparable to the English cord or fathom. See Warde, *Ecology, Economy and State Formation.*

49. LHAPotsdam, Rep. 3, Nr. 4884, 44–45. Copy of a letter to Frederick II, which was forwarded to the Kammer in Küstrin.

50. Ibid., 45. Signed in Bernau, 2 April 1766.

51. Ibid., 54. Bog iron ore was discovered in Torgelow, a town in Mecklenburg-Vorpommern, during the early eighteenth century. In 1753, a royal Prussian ironworks was established there.

52. Ibid., 49, 62–70, 76.

53. For the classic account, see Carlyle, *Friedrich II,* 5:120–40.

54. LHAPotsdam, Rep. 3, Nr. 4884, 102–115. One centner was roughly equivalent to a hundredweight, or 112 pounds, so the total amount of raw iron on hand was about 125,000 pounds.

55. LHAPotsdam, Rep. 3, Nr. 4884, 132–149.

56. Ibid., 102–10, 169.

57. Ibid., 184–85. Justi seems to have gotten their names, Kinell and Matthias, wrong. Cf. ibid., 169.

58. See chapter 3 on Göttingen.

59. LHAPotsdam, Rep. 3, Nr. 4884, 185–89. Marginal notes scribbled on Justi's letter suggest that the Kammer tried to move the weavers to a different house in nearby Drewitz.

60. LHAPotsdam, Rep. 3, Nr. 4884, 196, 209.

61. LHAPotsdam, Rep. 3, Nr. 4884, 217.

62. Ibid., 193, 239, 244.

63. Ibid., 244, 247. September 1766. The request was granted by Berlin in October.

64. Ibid., 264.

65. Ibid., 295–97. Copy of a letter from Justi to the Neumärkische Kammer. Original dated 21 July 1766.

66. Ibid., 273. As it turned out, the second blast furnace was plagued by delays and problems, for which Justi blamed the building inspector, Noach.

67. Bratring, *Statistisch-topographische Beschreibung,* 3:142–145; LHAPotsdam, Rep. 3, Nr. 4971, 8–13.

68. LHAPotsdam, Rep. 3, Nr. 4884, 287–91.

69. Justi, *Finanzwesens,* 237–40.

70. See Carlowitz, *Sylvicultura,* 78–125.

71. Justi, *Finanzwesens,* 240.

72. Ibid., 240–41.

73. Controller Wagner confirmed that Justi routinely ignored orders from the district forest administration.

74. LHAPotsdam, Rep. 3, Nr. 4884, 287–88.

75. Kleyensteuber's plan was in keeping with the "principles" of forestry as articulated by authorities like Carlowitz, (e.g., see *Sylvicultura,* 201–12).

76. LHAPotsdam, Rep. 3, Nr. 4884, 305–7. Appears to be from a special commission reporting directly to the king. The commission includes Baumeister Noach and some members of the Neumärkische Kammer.

77. Ibid., 290. Frederick II to Justi, 10 November 1766 (copy).

78. Ibid., 328–29. Cf. Carlowitz, *Sylvicultura,* 202–12.

79. For the view from Berlin on Justi's fall, see GStAPK, I HA Rep. 96B, Nr. 68, Bd. 68–70. "Minuten 1761–1766, Fragmente," and "Minuten 1767–1768" (Abschriften); Frensdorff, "Das Leben," 449–49; see also *Acta Borussica,* 14:277, 443.

80. Draft copy of a letter dictated to Schlabrendorff by Frederick II, 12 September 1766. GStAPK, I HA Rep. 96B, Nr. 68, Bd. 68, 502.

81. Anon. [H. K.], "O du mein Kupferberg," 211.

82. Draft of a letter from Frederick to Justi dated 19 February 1767, Potsdam. GStAPK, I HA Rep. 96B, Nr. 68, Bd. 70 (1767–1768), 66.

83. GStAPK, I HA Rep. 96B, Nr. 68, Bd. 70 (1767–1768), 66.

84. Ibid., 185.

85. Ibid., 208, 275.

86. Ibid., 275.

87. Cf. Johnson, *Frederick the Great and His Officials,* 214–16; *Acta Borussica,* 14:491–93, 527–32.

88. His report is in LHAPotsdam, Rep. 3, Nr. 4971, 7–36.

89. Ibid., 7–8.

90. Ibid., 1–6. This entire cabinet order, signed by Hagen, came down on 11 September 1768.

91. Ibid., 8–13.

92. Ibid., 8–23.

93. Tweedale, "Metallurgy and Technological Change," 190–91.

94. Smith, "The Discovery of Carbon in Steel," 154–55; Justi, *Chymische Schriften,* 2:107–8; LHAPotsdam, Rep. 3, Nr. 4971, 23–25.

95. See Minchinton, *Tinplate,* 1–24.

96. On the importance of skills for early modern manufacture and industry, see Rosenband, *Papermaking,* 97–101. Also, on the tradition of Harz iron and tinplate workers, see Beck, *Geschichte des Eisens,* 3:347–54.

97. LHAPotsdam, Rep. 3, Nr. 4971, 23–25, 25–27.

98. Ibid., 1–6.

99. Swedish iron was being underpriced by Siberian iron by this point, but it was still highly valued for its quality. See Florén and Rydén, "Mines, Furnaces and Forges," 71–90.

100. LHAPotsdam, Rep. 3, Nr. 4971, 1–6.

101. Johnson, *Frederick the Great and His Officials;* Weber, *Heynitz.*

102. LHAPotsdam, Rep. 3, Nr. 4983, 14–23.

103. Ibid.

104. Johnson, *Frederick the Great and His Officials,* 211.

105. Weber, *Heynitz,* 189–91; Fechner, "Schlesichen Berg- und Hüttenwesens," 145.

106. Johnson, *Frederick the Great and His Officials,* 228–30, 234–35. On this point, see also Sommer, *Die österreichischen Kameralisten,* 2: 226–34; Klein, "Johann Heinrich Gottlob Justi," 145–202; Johnson, *Frederick the Great and His Officials;* Harnack, *Akademie der Wissenschaften.*

107. For a brief recent review of this consensus, see Adam, *J. H. G. Justi,* 46–48.

108. LHAPotsdam, Rep. 3, Nr. 4998, 5–8.

CHAPTER FIVE

1. Anon., "Vorrede," *Oekonomische Weisheit und Thorheit,* vi–vii.

2. Anon., "Schreiben des Kantors in L** an den Verfasser A** zu ** über das Studium der ökonomischen und Cammeralwissenschaften," *Oekonomische Weisheit und Thorheit,* first part (1789), 1–46. Ludwig Heinrich von Jakob attributed the article to a "Geheimer Rath Barkhausen" in 1819; see Jakob, *Einleitung in das Studium der Staatswissenschaften.* I have, however, been unable to confirm it.

3. Anon., "Schreiben des Kantors," 6–7. Subsequent quotations in the text narrative are from this source.

4. On the education of scribes and *Amtmänner,* see McNeely, *Emancipation of Writing,* 35–66.

5. Beckmann coined the term *Technologie* and published the first textbook on the subject, *Anleitung zur Technologie,* in 1777. The fictitious letter, addressed as it was to the "author of *A** zu ***" and concerned with the new science of "*Technologie,*" was clearly aimed at Beckmann and the University of Göttingen.

6. The town was later named Kaiserslautern.

7. Tribe, *Governing Economy,* 92.

8. Lowood, *Patriotism, Profit, and the Promotion of Science,* 312.

9. Ibid., 313.

10. Tribe, *Governing Economy,* 92–94; Lowood, *Patriotism, Profit, and the Promotion of Science,* 313–14.

11. Plettenberg, *Hohe-Kameral-Schule,* 111–12, 129–30; Poller, *Schicksal,* 23–28, 30–38; Jung-Stilling, *Lebensgeschichte,* 447–54.

12. Jung-Stilling, *Autobiography,* 100 (here and elsewhere, I have emended Jackson's translation as necessary); *Lebensgeschichte,* 449–50.

13. Lautern was one of the nineteen primary districts (*Oberämter*) and forty-six towns in the lands of the Palatinate. The *Oberamt* Lautern, one of the poorest districts in the territory, included the so-called "Palatine Siberia," a largely barren area of swamps, pine forests, and outcroppings of bare rock. See Plettenberg, *Hohe-Kameral-Schule,* 69; Widder, *Beschreibung der kurfürstl. Pfalz am Rheine,* 4:165–70; Zink, "Geschichte der pfälzischen Landwirthschaft," 5–6.

14. Jung-Stilling, *Autobiography,* 105; and *Lebensgeschichte,* 469.

15. Tribe, *Governing Economy,* 99; and "Kameral Hohe Schule zu Lautern," 166–67. On Riem, see Poller, *Schicksal,* 13–16.

16. On the demographics of the *Gesellschaftsbewegung,* see Lowood, *Patriotism, Profit, and the Promotion of Science,* 68–73.

17. On Medicus, see Poller, *Schicksal,* 17–20.

18. Many of these are collected in Medicus, *Kleine ökonomische Aufsätze.*

19. Ibid., preface.

20. Medicus, *Kleine ökonomische Aufsätze,* 65, 85.

21. Medicus had ties with prominent Swiss and German physiocrats, among them Riem, Karl Friedrich of Baden, Isaak Iselin, and Johann August Schlettwein.

22. Medicus, *Kleine ökonomische Aufsätze,* 87.

23. The *Landkalender* cajoled and persuaded, rather than dictating change. Nevertheless, like Justi's *Policey-Amts-Nachrichten,* its form and function resembled police ordinances.

24. Häberle, "Siegelbach," 44; Plettenberg, *Hohe-Kameral-Schule,* 90–95.

25. Each "hectare" was probably 1.3 times the size of a present-day hectare ($10{,}000m^2$). See Plettenberg, *Hohe-Kameral-Schule,* 63.

26. Häberle, "Der botanische Garten zu Kaiserslautern," 44–45.

27. Anon. "Bericht über Siegelbach," 13–14.

28. Jung-Stilling, *Lebensgeschichte,* 472.

29. Ibid.,

30. Müller, *Geschichte des höheren Schulwesens,* 14–15.

31. Jung-Stilling, *Lebensgeschichte,* 485–86.

32. Häberle, "Siegelbach," 45–46; Plettenberg, *Hohe-Kameral-Schule,* 93–95.

33. Cited in Häberle, "Siegelbach," 46; and in Plettenberg, *Hohe-Kameral-Schule,* 94–95.

34. Heidelberg, home to the oldest university in the German lands of the empire, had been capital of the *Kurpfalz* since 1583, when Karl Philipp decided to move. See Haas, *Pfalz am Rhein,* 179; Schnabel, *Abhandlungen und Vorträge,* 65–66.

35. Schnabel, *Abhandlungen und Vorträge,* 74.

36. Haas, *Pfalz am Rhein,* 151–59.

37. The most famous and successful of these enterprises was the porcelain factory. Frankenthaler porcelain flourished until the French Revolution. See Haas, *Pfalz am Rhein,* 181.

38. Plettenberg, *Hohe-Kameral-Schule,* 96.

39. GLAKarlsruhe, Abt. 205, Nr. 1112; Müller, *Geschichte des höheren Schulwesens,* 10–15.

40. Lautern's so-called siamoise was a kind of *Zitse,* or colored calico. Most European calico manufactories printed the pattern and primary color on the material. The remaining colors would be painted in by hand. According to Justi, it was primarily the number of colors that distinguished the so-called *Zitse* from other kinds of calico (*Cattune*). "The colored calicos are actually called *Zitse,* or better *Chitse,* even though these names actually cover every calico with three or more colors. . . . Printed calicos are either those that have more than two colors, which, as mentioned earlier, have the name *Zitse;* or they are composed of two colors, in which case they are then specifically called calicos" (Justi, *Manufacturen und Fabriken,* 2:116–18).

41. Copy of the electoral patent for a linen manufactory in Lautern; issued by the Elector Palatine to the Lautern Physical-Œconomic Society, 18 April 1771. GLA-Karlsruhe, Abt. 205, Nr. 1112 (hereafter referred to as "1771 Patent").

42. 1771 Patent, secs. 1–7, 11, and 12.

43. Webler, *Kameral-Hohe-Schule,* 40; Plettenberg, *Hohe-Kameral-Schule,* 97–99.

44. Plettenberg, *Hohe-Kameral-Schule,* 97.

45. Jung-Stilling, *Lebensgeschichte,* 440–41.

46. Jung-Stilling's essays were well received, and he was eventually named an "external member" of the Physical-Œconomic Society.

47. Webler, *Kameral-Hohe-Schule.*

48. Müller, *Geschichte des höheren Schulwesens,* 14; Häberle, "Siegelbach," 44.

49. Webler, *Kameral-Hohe-Schule,* 48; Müller, *Geschichte des höheren Schulwesens,* 16–18; Plettenberg, *Hohe-Kameral-Schule,* 97.

50. Tribe, "Kameral Hohe Schule zu Lautern," 173–74.

51. Stieda, "Hohen Kameralschule in Kaiserslautern," 340–53.

52. *Bemerkungen der Kuhrpfälzischen Physikalisch-Ökonomischen Gesellschaft* (1771–85).

53. Ordinance promulgated by the Palatine Hofkammer on 19 December 1778. A copy of the printed ordinance was included among the various petitions and supplications to the ministry for special exception from the regulation. See GLAKarlsruhe, Abt. 25, Nr. 1110.

54. The ordinance referred to "Kurfürstliche Kameral- und Administrations Ober- und Landbedienungen" and "Kameral-Obere Stellen und Landsbedienungen, wie zu Geistlicher Administrations-Rathsstellen und Unterbedienungen fähig seyn." In other words, it construed servants of the Kammer broadly.

55. Marc Anton Faßele, “adjungirter Collector von der Collektür in Mannheim,” to Hofkammer, 5 January 1780. GLAKarlsruhe, Abt. 25, Nr. 1112.

56. Ordinance promulgated by the Palatine Hofkammer on 6 July 1779. GLA-Karlsruhe, Abt. 25, Nr. 1110.

57. All of these cases and the faculty responses are collected in GLAKarlsruhe, Abt. 25, Nr. 1112, in the same bundle as the copy of the 1778 ordinance. The case of “Geistlicher Administrations Collector Gotthard zu Kaub,” which dragged on for over two years, proved particularly stubborn. In the end, however, Medicus prevailed here too. On 29 September 1784, after the cameral academy had moved to Heidelberg, the authorities (the order was signed by Oberndorff) ordered Gotthard to attend. GLAKarlsruhe, Abt. 25, Nr. 1112.

58. Professor Succow described the elector’s visit in the protocols of the Physical-Œconomic Society. His account is reprinted in Webler, *Kameral-Hohe-Schule,* 48, and cited in Plettenberg, *Hohe-Kameral-Schule,* 123.

59. GLAKarlsruhe, Abt. 205, Nr. 1115. See appendix 3 for a translation.

60. Keith Tribe has cited and reproduced Moshammer’s published plan at some length. See his *Governing Economy,* 95–96.

61. UniArchMü, E-II-214a. Moshammer’s dossier at the Ludwig Maximilian University of Munich includes various plans, among them an extensive and detailed 1799 plan for the “Cameral Institute” in Ingolstadt. This was, however, fifteen years after the 1784 plan was drafted. By then, Moshammer was an ordinary professor and more secure in his position. See UniArchMü, E-II-214b.

62. There are two sets of documents. The first set, partially damaged and difficult to read, was penned by Medicus in February of 1784. These documents include instructions for professors at the proposed faculty in Ingolstadt (10 February 1784). The second bears no date or signature, but it was also written by Medicus. He called this second document the “main plan.” See GLAKarlsruhe, Abt. 205, Nr. 1115. I have appended a translation (see appendix 3). Hereafter, I refer to the document as the “Plan for Ingolstadt.” Emil Müller, who cited part of the “Plan for Ingolstadt” in his history of Lautern, suggested that the document was written by Freiherr Christoph von Hautzenberg, commander of a regiment in Mannheim (though he was not sure). Other documents in the bundle—and these are actually signed by Medicus—refer to an attached “plan.” The documents signed by Medicus use the same terms and observe the same structure as the unsigned and attached “Plan for Ingolstadt.” Medicus’s published 1784 *Nachricht an das Publikum* contains whole phrases and expressions that appear in the “Plan for Ingolstadt.” For the claim that Hautzenberg authored the “Plan for Ingolstadt,” see Müller, *Geschichte des höheren Schulwesens,* 46–48.

63. Medicus signed and dated documents from the first set. They were written in Munich between 10 and 24 February 1784. See GLAKarlsruhe, Abt. 205, Nr. 1115.

64. Initially, Karl Theodor did not want to move to Munich; only the combined influence of Frederick II and Joseph II convinced him to leave Mannheim. See Rall, *Kurfürst Karl Theodor,* 219.

65. Holborn, *A History of Modern Germany,* vol. 2, 293–94.

66. Rall, *Kurfürst Karl Theodor,* 275–77; 181–82.

67. Medicus, *Nachricht an das Publikum,* 13; Jung-Stilling, *Lebensgeschichte,* 536–37.

68. Haas, *Pfalz am Rhein,* 131.

69. GLAKarlsruhe, Abt. 205, Nr. 1115, "Plan for Ingolstadt."

70. Medicus published some of the same thoughts in his 1784 *Nachricht an das Publikum,* but the system is spelled out in greater detail in his unpublished "Plan for Ingolstadt."

71. GLAKarlsruhe, Abt. 205, Nr. 1115, "Plan for Ingolstadt."

72. Ibid.,

73. Semer was a student in Lautern during 1781. In 1786, two years after Medicus (unsuccessfully) recommended him for Ingolstadt, Semer became extraordinary professor in Heidelberg. See Poller, *Schicksal,* 52–53; GLAKarlsruhe, Abt. 205, Nr. 1115, "Plan for Ingolstadt."

74. GLAKarlsruhe, Abt. 205, Nr. 1115, "Plan for Ingolstadt."

75. It was not unusual for the authorities to hear complaints from starving professors. See Clark, "Ministerial Archive," 430–36, 457–61.

76. Tribe, "Kameral Hohe Schule zu Lautern," 180–81; and *Governing Economy,* 110; Plettenberg, *Hohe-Kameral-Schule,* 97.

77. GLAKarlsruhe, Abt. 205, Nr. 1115, "Plan for Ingolstadt."

78. Ibid.,

79. Informants and spies were used regularly by territorial governments in the empire. On 18 December 1784, Oberndorff sent an excerpt of Wrede's monthly report to Medicus. See GLAKarlsruhe, Abt. 205, Nr. 1112.

80. Wrede to electoral authorities, GLAKarlsruhe, Abt. 205, Nr. 1112.

81. Jung-Stilling, *Autobiography,* 120; and *Lebensgeschichte,* 535–36.

82. Jung-Stilling calls it "[der] Saal der staatswirthschaftlichen hohen Schule." Jackson chose to translate this as "the hall of the statistical academy," which is misleading. Unfortunately, there is no good translation for the phrase. Nevertheless, Jung-Stilling's use of the adjective *staatswirthschaftlich* here would have called to mind Justi's *Staatswirthschaft,* still among the most popular of cameralist textbooks during the late eighteenth century. See Justi, *Staatswirthschaft,* vol. 1, preface.

83. Jung-Stilling, *Autobiography,* 120; and *Lebensgeschichte,* 535–36.

84. Jung-Stilling's years in Lautern were plagued by concerns over debts. In 1778, for example, he owed Medicus twenty-eight florins that he could not afford to repay. Jung-Stilling, *Lebensgeschichte,* 546–37.

85. Jung-Stilling, *Über den Geist der Staatswirthschaft,* 6.

86. Ibid., 7–8, 30.

87. Jung-Stilling had been regarded with deep suspicion in Lautern for his Pietist tendencies and found himself ostracized more than once for being a "Schwärmer," or enthusiast.

88. Jung-Stilling, *Über den Geist der Staatswirthschaft,* 12.

89. This explains, for example, Medicus's harsh appraisal of Moshammer in his "Plan for Ingolstadt." The Ingolstadt professor may have presented a curriculum that resembled the Lautern system—he even mentioned Lautern as a model—but as far as Medicus was concerned, it was all show. Moshammer had neither the ability nor the resources to teach the necessary "foundational" and "resource" sciences.

CHAPTER SIX

1. I would like to thank Hans-Jörg Ruge, the director of the Thüringische Staatsarchiv in Gotha, for his help with this and other annoying questions.

2. See the documents reproduced in Jacobsen and Ruge, eds., *Ernst der Fromme,* 301–22.

3. The chart itself is located in ThStAGotha, GA KK III, Nr. 1, 77. For evidence that it was hanging in the *Kammerstüblein*, see ThStAGotha, GA KK III, Nr. 14, 9.

4. Foucault, *Discipline and Punish,* 173.

5. Bödeker, "On the Origins of the 'Statistical Gaze,'" 169–71.

6. Though *Kammerrat* Christoph vom Hagen received some payments after 1756, he was effectively replaced by Seckendorff in that year. See ThStAGotha, GA, KK III, Nr. 3, Nr. 6, 10–11. There is some uncertainty about the overlapping careers and official positions of Hagen and Seckendorff, but it is clear that Seckendorff was, in practice, in charge of *Kammersachen* after April 1656. See Ruge, "Ubersicht über die Besoldung"; Heß, *Geheimer Rat,* 50–53.

7. There are two large *Kammerordnungen* in Gotha's secret archive: a copy of the 1656 *Ordnung* and another version from 1666 (revised by Hiob Ludolf, who replaced Seckendorff in 1664). ThStAGotha, GA, KK III, Nr. 5, Nr. 6.

8. Ibid., Nr. 3.

9. Ibid., Nr. 1, Nr. 3.

10. See Klinger, "Zur Staatsbildung Ernsts des Frommen."

11. ThStAGotha, GA, KK III, Nr. 6, 100.

12. Jacobsen, "Die Brüder Seckendorff," 117–19.

13. Tribe, *Governing Economy,* 10–17.

14. See Medicus's 1784 "Plan for Ingolstadt" (appendix 3) for the paradigmatic example.

15. There has been increasing recognition of the importance of the cameral sciences for the history of science; see Clark, *Academic Charisma;* Cooper, "Possibilities of the Land"; Koerner, *Linnaeus*. I am pushing the argument a little farther; that is, many of the cameral sciences *were* natural sciences.

16. See appendix 3.

17. Novick's *Noble Dream* is still the best account.

18. For the anti-skeptical stance, see Appleby, Hunt, and Jacob, *Telling the Truth about History,* 308.

19. Cf. Sabean, *Kinship in Neckarhausen;* Medick, *Weben und Überleben.*

20. On the limitations of *Alltagsgeschichte,* see Gregory, "Is Small Beautiful?"

21. See Gorski, "Disciplinary Revolution and State Formation," 303–7.

22. See Bödeker, "On the Origins of the 'Statistical Gaze,'" 176–82.

23. For a recent example, see Blackbourn, *The Conquest of Nature.*

24. See Miller, "Nazis and Neo-Stoics."

25. See Jacob and Stewart, *Practical Matter.*

26. Small, *The Cameralists,* 594–95.

27. Foucault, *Discipline and Punish,* 154.

28. Weber, *Wirtschaft und Gesellschaft,* 354–55; and *The Protestant Ethic,* 102–25.

29. Oestreich, *Neostoicism and the Early Modern State.*

30. Elias, *The Civilizing Process.*

31. For the argument in its purest form, see Gorski, *Disciplinary Revolution,* xv–xvii, 1–38. See also Schilling, *Kirchenzucht und Sozialdisziplinierung:* Johnson and Monkkonen, eds., *The Civilization of Crime;* Finzsch and Jütte, eds., *Institutions of Confinement.* For a history of science perspective, see the essays in Becker and Clark, eds., *Little Tools of Knowledge.*

32. See the thousands of texts catalogued in Humpert, *Bibliographie der Kameralwissenschaften* (1937).

APPENDIX THREE

GLAKarlsruhe, Abt. 205, Nr. 1115. Reproduced with the permission of the Generallandesarchiv Karlsruhe. My translation. All emphases (i.e., italics) are in the original.

1. Franz Xaver Moshammer (1756–1826), lecturer at Ingolstadt.

2. Medicus is referring here to the French physiocrats, who considered agriculture the basis of national wealth.

3. Engelbert Martin Semer was a student in Lautern during 1781. In 1786, two years after Medicus (unsuccessfully) recommended him for Ingolstadt, Semer became extraordinary professor in Heidelberg.

4. I was not able to find the "Special Order" to which Medicus here refers.

BIBLIOGRAPHY

ARCHIVAL SOURCES

Archiv der Göttingen Akademie der Wissenschaften (AkArchGö)

AkArchGö, Pers 17, 3–7; and AkArchGö, Pers 30, 3, 47 (Justi's appointment to Göttingen's Royal Society).

AkArchGö, Etat 6, 2, 1–3 (Justi's suggestions for improving the income of Göttingen's Royal Society).

AkArchGö, Scient 12, 21 (Regarding Justi's manuscript: "De metallorum per colores, quos in vitro producunt, arte probatoria").

AkArchGö, Pers 12, 6 and 9; and AkArchGö, Pers 30, 3, 114 (Johann Beckmann's appointment to Göttingen's Royal Society).

AkArchGö, Scient 195, 9, 1 and 10; AkArchGö, Scient 196, Fasz. 3, 13 and 19 (Concerning the prize question on leather tanneries).

Archiwum Państwowe we Wrocławiu (ArchBreslau)

ArchBreslau, Sygn. 00303, Nr. 406 (Concerning the recruitment of mining and smelting students, 1770–1833).

Brandenburgisches Landeshauptarchiv Potsdam (LHAPotsdam)

LHAPotsdam, Rep. 3, Nr. 4884 (Documents relating to the ironworks in Vietz and Kutzdorff).

LHAPotsdam, Rep. 3, Nr. 4971 (Report on the combined iron works in the Neumark).

LHAPotsdam, Rep. 3, Nr. 4982 (Justi's memorandum to the General Directory on the Prussian ironworks).

LHAPotsdam, Rep. 3, Nr. 4983 (Includes secret memorandum from Kriegs-Rat Jaeckel).

LHAPotsdam, Rep. 3, Nr. 4998 (Includes the results of an investigation conducted by the Mining and Smelting Administration in 1800).

Forschungsbibliothek Gotha (FBGotha)

FBGotha, Handschriften, Chart. A 918, Nr. 84 (Correspondence, Seckendorff).
FBGotha, Handschriften, Chart. B 1918 II (Happe, Gottlob Christian).

Geheimes Staatsarchiv Preußischer Kulturbesitz, Berlin (GStAPK)

GStAPK, I HA Rep. 121, Abt. D, Tit. II, Sect. 1, Nr. 101, Bd. 1–2 (Concerning the instruction of mining, smelting, and salt mining candidates, 1770–71).
GStAPK, I HA Rep. 96B, Nr. 68, Bd. 68 (Abschriften) (Minutes and fragments, 1761–66).
GStAPK, I HA Rep. 96B, Nr. 68 Bd. 70 (Abschriften) (Minutes and fragments, 1767–68).Geheimes Staatsarchiv Preußischer Kulturbesitz, Berlin (GStAPK)

Generallandesarchiv Karlsruhe (GLAKarlsruhe)

GLAKarlsruhe, Abt. 205, Nrn. 1110–15 (The Kameral-Hohe-Schule in Lautern and its transfer to Heidelberg).
GLAKarlsruhe, Abt. 205, Nr. 1136 (Concerning the rules of conduct for students at the Kameral-Hohe-Schule).

Niedersächsiches Hauptstaatsarchiv Hannover (HStArchHann)

HStArchHann, Hann. 80 Hildesheim I, F, Nr. 109 (Grätzel's manufactory in Göttingen).
HStArchHann, Dep. 7 B, Nr. 2026 (Foundation of the University of Göttingen, 1734–37).
HStArchHann, Dep. 7 B, Nrn. 2037–41 (Documents concerning Celle's Zucht- und Tollhaus).
HStArchHann, Dep. 7 B, Nrn. 461–66 (Zuchthaus accounts).
HStArchHann, Dep. 7 B, Nrn. 2042–46 (Contributions to maintain the *Zuchthaus* [1748–1777]).

Niedersächsische Staats- und Universitätsbibliothek Göttingen, Handschriften Abteilung (HandAbtUniGö)

HandAbtUniGö, Bibl. Arch., Manuale. (Handwritten acquisition catalogues of the old Göttingen university library.)

HandAbtUniGö, 4 Cod., Philos. 149, Münchhausen, Briefe
HandAbtUniGö, 2 Cod. MS Michael. 324

Oberbergamt Clausthal-Zellerfeld (OBAClausthal)

OBAClausthal, Fach 761/27 and Fach 762/29 (Leibniz's windmill project in the Harz mines ["Windmühlenkünste"]).

Sächsisches Hauptstaatsarchiv Dresden (StADresden)

StADresden, Loc. 514 (The establishment of a mining academy in Freiberg).
StADresden, Loc. 1327 (The direction of mines under Heynitz, 1765–74).
StADresden, Loc. 36216 (Proposals made for the benefit of the mining academy in Freiberg).

Stadtarchiv Göttingen (StadtArchGö)

StadtArchGö, AA, Polizeyverwaltung, Pol 1, 2 ,1 (Göttingen police records, 1750–55).
StadtArchGö, AA, Gewerbesachen, Manufaktur-Gericht, Nr. 49 (The establishment of a manufactures court in Göttingen, 1743–1816).
StadtArchGö, AA, Industrie, Fabrik- und Manufaktursachen, Nr. 15 (The improvement of manufactures, 1724–25).
StadtArchGö, AA, Industrie, Fabrik- und Manufaktursachen, Nr. 17 (Manufactories for striped and colored linen).
StadtArchGö, AA, Abgaben, Onera Nr. 58 (The taxes and debts of Commerce Commissar Grätzel).

Thüringisches Staatsarchiv Altenburg (ThStAAltenburg)

ThStAAltenburg, Nr. 1066 ("Domestica des Veit Ludwig von Seckendorff").

Thüringisches Staatsarchiv Gotha (ThStAGotha)

ThStAGotha, Kammer Immediate, Nr. 1201 (As an experiment, ten "mast trees" are taken from the Thüringerwald, floated to Bremen, and then sent farther along to Amsterdam).
ThStAGotha, Kammersachen Insgemein, Nr. 1041–42 (Descriptions of the Thüringerwald by district).
ThStAGotha, GA KK III, Nrn. 1–14 (Kammerordnung).
ThStAGotha, Kammer Immediate, Nr. 1660k (Protocols from the Kammer in Friedenstein, 1683–84).

ThStAGotha, StAGotha, GA, U U I, 1 (Veit Ludwig von Seckendorff).
ThStAGotha, E IV §, Nrn. 2a and 5 (Seckendorff's trips with the princes).

Universitätsarchiv der TU Bergakademie Freiberg (UniArchFreiberg)

UniArchFreiberg, Akte OBA 7917, vol. 1.
UniArchFreiberg, Akte OBAJ 6695 (Main plan of the *Bergakademie*).
UniArchFreiberg, Akte OBAJ 6706 (Warning for the cadets who hauled ore out of the mines without permission).
UniArchFreiberg, Akte OBAJ 8547 (Preliminary establishment of the conferences proposed by Heynitz).
UniArchFreiberg, Akte OBAJ 7762 (Increase of funds through the *obergebirgische* silver yields).
UniArchFreiberg, Sekt. 91d, Acta 8045 (Documents concerning the Freiberg cadets from the consistory in Dresden).

Universitätsarchiv Göttingen (UniArchGö)

UniArchGö, Kur 1.38 (Published report on the condition of the university).
UniArchGö, Kur 1.67–68 (Decree about students expelled from other universities).
UniArchGö, Kur 3.k.10–12, 20 (Assorted crimes and punishments).
UniArchGö, Kur 3.1.108 (The dispute between professors Beckmann and Kästner).
UniArchGö, Kur 4.IV.b.13 (Johann Andreas Murray, 1763–91).
UniArchGö, Kur 4.IV.b.21 (Johann Friedrich Gmelin, 1774–94).
UniArchGö, Kur 4.V.a.7 (The attempt by the philosophy faculty to hold botanical lectures, 1767).
UniArchGö, Kur 4.V.b.14 (Gottfried Achenwall, 1748–73).
UniArchGö, Kur 4.V.b.24 (Abraham G. Kästner, 1755–1801).
UniArchGö, Kur 4.V.b.35 (Johann. C. E. Springer, 1765–97)
UniArchGö, Kur 4.V.b.36 (Johann Beckmann, 1766–1811).
UniArchGö, Kur 4.V.g.3 (The purchase of Grätzel's *Naturalien-Cabinett,* 1773).
UniArchGö, Kur 4.V.g.1 (The suggested purchase of a *Naturalien-Cabinett,* 1756).
UniArchGö, Kur 4.V.i.14 (The purchase of Hofrath Beckmann's model collection).
UniArchGö, Kur 4.V.1.1 (Establishment of the œconomic garden, 1767).
UniArchGö, Kur 5.c.2 (Investigation of police shortcomings, 1735–36).
UniArchGö, Kur 5.c.7–15 (Police commission records, 1753–71).

Universitätsarchiv München (UniArchMü)

UniArchMü, E-II-214a (Franz Xaver Moshammer, 1786–1824).
UniArchMü, E-II-214b (Friedrich von Moshammer).

PRINTED SOURCES

Abbot, Andrew. *The System of Professions: An Essay on the Division of Expert Labor.* Chicago: University of Chicago Press, 1988.

Achenwall, Gottfried. *Abriß der neuesten Staatswissenschaft der vornehmsten Europäischen Reiche und Republicken.* Göttingen: Schmidt, 1749.

———. *Staatsverfassung der heutigen vornehmsten europäischen Reiche.* Göttingen: Vandenhoeck, 1756.

———. *Die Staatsklugheit nach ihren ersten Grundsätzen.* Göttingen: Vandenhoeck, 1761.

Adam, Ulrich. *The Political Economy of J. H. G. Justi.* Oxford: Lang, 2006.

Ågren, Maria, ed. *Iron-Making Societies: Early Industrial Development in Sweden and Russia, 1600–1900.* Providence, RI: Berghahn, 1998.

———. "Introduction: Swedish and Russian Iron-Making As Forms of Early Industry." In Ågren, *Iron-Making Societies,* 3–32.

Albrecht-Birkner, Veronika. *Reformation des Lebens: Die Reformen Herzog Ernsts des Frommen von Sachsen-Gotha und ihre Auswirkungen auf Frömmigkeit, Schule und Alltag im ländlichen Raum (1640–1675).* Leipzig: Evangelische Verlagsanstalt, 2002.

Alder, Ken. *Engineering the Revolution: Arms and Enlightenment in France, 1763–1815.* Princeton, NJ: Princeton University Press, 1997.

Allgemeine deutsche Biographie. 56 vols. Leipzig: Duncker & Humblot, 1875–1912.

Altmann, Götz. *Erzgebirgisches Eisen: Geschichte—Technik—Kultur.* Dresden: Sächsisches Druck und Verlagshaus, 1999.

Anon. "Bericht über Siegelbach." In *Bemerkungen der Kuhrpfälzischen physikalisch-ökonomischen Gesellschaft vom Jahre 1777,* 13–15. Mannheim and Lautern: Hof- und Akademischen Buchhandlung, 1779.

———. "Brief eines Reisenden über Zelle." *Journal von und für Deutschland* 10 (1786): 351–56.

———. *Erneuerte und durch Anordnung eines besondern Arbeitshauses verbesserte Armenordnung der Stadt Celle und deren Vorstädte. 7 Nov. 1783.* Hannover: n.p., 1783.

———. [Teutophylus, Christianus]. *Geprüffte Gold-Grube der Universal-Accise.* Dresden: Günther, 1687.

———. [H. K.] "O du mein Kupferberg." *Schlesiche Bergwacht* 12 (1957): 211–12.

———. *Der Nutzen und die Nutzungen des Bergwerks.* Frankfurt and Leipzig: n.p., 1766.

———. "Schreiben des Kantors in L** an den Verfasser A** zu ** über das Studium der ökonomischen und Cammeralwissenschaften." *Oekonomische Weisheit und Thorheit* 1 (1789): 1–46.

———. *Ueber das Studium der ökonomischen und Cameral-Wissenschaften.* Erfurt: Keyser, 1789.

———. *Der Ungetreue Rechnungs-Beambte.* Eisenach: Thilo, 1684.
———. "Vorrede." *Oekonomische Weisheit und Thorheit* 1 (1789): iii–xiv.
———. *Zur Hundertjährgen Gedächtniß-Feyer der Grätzelschen Fabrik in Göttingen.* Göttingen: n.p., 1814.
Appleby, Joyce, Lynn Hunt, and Margaret Jacob. *Telling the Truth about History.* New York: W. W. Norton, 1994.
Aretin, Karl Otmar von, ed. *Der Aufgeklärte Absolutismus.* Cologne: Kiepenheuer & Witsch, 1974.
———. *Heiliges Römisches Reich 1776 bis 1806: Reichsverfassung und Staatssouveränität.* 2 vols. Wiesbaden: Steiner, 1967.
Arndt, Adolf. *Zur Geschichte und Theorie des Bergregals und der Bergbaufreiheit.* Halle: Pfeffer, 1879.
Ashworth, William J. *Customs and Excise: Trade, Production, and Consumption in England, 1640–1845.* New York: Oxford University Press, 2003.
———. "The Ghost of Rostow: Science, Culture and the British Industrial Revolution." *History of Science* 46, no. 3 (2008): 249–74.
———. "The Intersection of Industry and the State in Eighteenth-Century Britain." In *The Mindful Hand: Inquiry and Invention from the Late Renaissance to Early Industrialisation,* edited by Lissa Roberts, Simon Schaffer, and Peter Dear, 349–77. Amsterdam: Royal Netherlands Academy of Sciences, 2007.
Assman, Klaus. "Verlag-Manufaktur-Fabrik: Die Entwicklung großbetrieblicher Unternehmensformen im Göttinger Tuchmachergewerbe." In *Handwerksgeschichte in neuer Sicht,* edited by Wilhelm Abel, 211–39. Göttingen: Schwarz, 1978.
Baker, Keith. *Condorcet: From Natural Philosophy to Social Mathematics.* Chicago: University of Chicago Press, 1975.
Bärens, Johann Georg. "Kurtze Nachricht von Göttingen." In "Ein Bericht über Göttingen, Stadt und Universität, aus dem Jahre 1754," edited and reprinted by Ferdinand Frensdorff. *Jahrbuch des Geschichtvereins für Göttingen und Umgebung* 1 (1908): 43–115.
Bartels, Christoph. *Vom frühneuzeitlichen Montangewerbe zur Bergbauindustrie: Erzbergbau im Oberharz, 1635–1866.* Bochum: Deutsches Bergbau-Museum, 1992.
Baumgärtel, Hans. "Alexander von Humboldt und der Bergbau." In *Alexander von Humboldt (1769–1859): Seine Bedeutung für den Bergbau und die Naturforschung.* Freiberger Forschungshefte, series D, no. 33, 115–49. Berlin: Akademie Verlag, 1960.
———. *Bergbau und Absolutismus: Der sächsische Bergbau in der zweiten Hälfte des 18. Jahrhunderts und Massnahmen zu seiner Verbesserung nach dem Siebenjährigen Kriege.* Freiberger Forschungshefte, series D, no. 44. Leipzig: Deutscher Verlag für Grundstoffindustrie, 1963.
———. *Vom Bergbüchlein zur Bergakademie: Zur Entstehung der Bergbauwissen-*

schaften zwischen 1500 und 1765/1770. Freiberger Forschungshefte, series D, no. 50. Leipzig: Deutscher Verlag für Grundstoffindustrie, 1965.

Becher, Johann Joachim. *Politische Discurs.* 1688. Reprint, Glashütten: Auvermann, 1972.

Beck, August. *Ernst der Fromme, Herzog zu Sachsen-Gotha und Altenburg: Ein Beitrag zur Geschichte des 17. Jahrhunderts.* 2 vols. Weimar: Böhlau, 1865.

———. *Geschichte der Stadt Gotha.* Gotha: Thienemann, 1870.

———. *Geschichte des gothaischen Landes.* Gotha: Thienemann, 1860.

Beck, Herrmann. *The Origins of the Authoritarian Welfare State in Prussia.* Ann Arbor: University of Michigan Press, 1997.

Beck, Ludwig. *Die Geschichte des Eisens in technischer und kulturgeschichtlicher Beziehung.* 5 vols. Braunschweig: Vieweg und Sohn, 1891–99.

Becker, Peter, and William Clark, eds. *Little Tools of Knowledge: Historical Essays on Academic and Bureaucratic Practices.* Ann Arbor: University of Michigan Press, 2001.

Beckert, Manfred. *Johann Beckmann.* Leipzig: Teubner, 1983.

Beckmann, Johann. *Anfangsgründe der Naturhistorie.* Göttingen and Bremen: Förster, 1767.

———. *Anleitung zur Handlungswissenschaft.* Göttingen: Vandenhoeck & Ruprecht, 1789.

———. *Anleitung zur Technologie.* Göttingen: Vandenhoeck, 1777.

———. *De historia naturali veterum libellus primus.* Göttingen and St. Petersburg: Dieterich, 1766.

———. *Gedanken von der Einrichtung ökonomischer Vorlesungen.* Göttingen: Vandenhoeck, 1767.

———. *Grundsätze der teutschen Landwirthschaft.* Göttingen: Dieterich, 1769.

———. ed. *Physikalisch-oekonomische Bibliothek.* 23 vols. Göttingen: Vandenhoeck, 1770–1806.

———. *Schwedische Reise nach dem Tagebuch der Jahre 1765–1766.* Edited by Thore Magnus Fries. Lengwil: Libelle, 1995.

———. *Vorrath kleiner Anmerkungen ueber mancherley gelehrte Gegenstände.* 3 vols. Göttingen: Röwer, 1795–1806.

Beckmann, Johann, and Johann Bergius, eds. *Sammlung auserlesener teutscher Landesgesetze, welche das Policey- u: Cameralwesen zum Gegenstande haben.* Frankfurt: n.p., 1781.

Bérenger, Jean. *A History of the Habsburg Empire, 1700–1918.* Translated by C. A. Simpson. London: Longman, 1997.

Bicker, Friedrich. "Das Staatsschuldenproblem in der Lehre der Cameralistik. (Seckendorff, Justi, Sonnenfels)." PhD diss., Gießen University, 1928.

Bischoff, Johann N., ed. *Versuch einer Geschichte der Färberkunst von ihrer Entstehung an bis auf unsere Zeiten entworfen und mit einer Vorrede von Johann Beckmann.* Stendal: n.p., 1780.

Biskup, Thomas. "The University of Göttingen and the Personal Union." In Simms and Riotte, *Hanoverian Dimension,* 128–60.

Black, Jeremy. "Hanoverian Nexus: Walpole and the Electorate." In Simms and Riotte, *Hanoverian Dimension,* 10–27.

Blackbourn, David. *The Conquest of Nature: Water, Landscape, and the Making of Modern Germany.* New York: W. W. Norton, 2006.

Blackbourn, David, and Geoff Eley. *Peculiarities of German History: Bourgeois Society and Politics in Nineteenth-Century Germany.* Oxford: Oxford University Press, 1984.

Bleek, Wilhelm. *Von der Kameralausbildung zum Juristenprivileg.* Berlin: Colloquium, 1972.

Blumenbach, Johann Friedrich. *Handbuch der Naturgeschichte.* Gottingen: Dieterich, 1779.

Bödeker, Hans Erich. "On the Origins of the 'Statistical Gaze': Modes of Perception, Forms of Knowledge and Ways of Writing in the Early Social Sciences." In Becker and Clark, *Little Tools of Knowledge,* 169–95.

———. "Prozesse und Strukturen politischer Bewusstseinbildung der deutschen Aufklärung." In *Aufklärung als Politisierung—Politisierung als Aufkärung,* edited by Hans Erich Bödeker and Ulrich Herrmann, 143–62. Hamburg: Meiner, 1987.

———. "Das staatswissenschaftliche Fächersystem im 18. Jahrhundert." In *Wissenschaft im Zeitalter der Aufklärung,* edited by Hans Erich Bödeker et al., 143–62. Göttingen: Vandenhoeck & Ruprecht, 1985.

Bog, Ingomar. *Der Reichsmerkantilismus.* Stuttgart: Fischer, 1959.

Bohnsack, Almut. *Spinnen und Weben: Entwicklung von Technik und Arbeit im Textilgewerbe.* Reinbeck bei Hamburg: Rowohlt, 1981.

Böll, Friedrich Phillip Carl. "Bemerkungen über Johann Jacob Mosers Rede, wie Universitäten, besonders in der juristischen Facultät, in Aufnahme zu bringen und darinn zu enthalten." In Rössler, *Gründung,* 468–86.

———. *Sendschreiben über die Anfrage, in was für einem Zustand sich die Rechtsgelehrsamkeit auf der blühenden Georg Augusta befinde.* Colmar: Neukirch, 1775.

———. *Das Universitätswesen in Briefen.* N.p., 1782.

Bonney, Richard, ed. *The Rise of the Fiscal State in Europe,* c. *1200–1815.* Oxford: Oxford University Press, 1999.

Böttiger, C. W. *Geschichte des Kurstaates und Königreiches Sachsen.* 4 vols. Gotha: Perthes, 1867–73.

Bowler, Richard Carl. "Bildung, Bureaucracy, and Political Economy: Karl Heinrich Rau and the Development of German Economics." PhD diss., University of California, Los Angeles, 1996.

Bozorgnia, S. M. H. *The Role of Precious Metals in European Economic Development.* Westport, CT: Greenwood, 1988.

Brandes, Ernst. *Ueber den gegenwärtigen Zustand der Universität Göttingen.* Göttingen: n.p., 1802.

Bratring, Friedrich Wilhelm August. *Statistisch-topographische Beschreibung der gesammten Mark Brandenburg: für Statistiker, Geschäftsmänner, besonders für Kameralisten.* 3 vols. Berlin: Maurer, 1804–9.

Breuilly, John. "Hamburg: The German City of Laissez-Faire." *Historical Journal* 35, no. 3 (1992): 701–12.

Brewer, John, and Eckhart Hellmuth, eds. *Rethinking Leviathan: The Eighteenth-Century State in Britain and Germany.* Oxford: Oxford University Press, 1999.

Broman, Thomas. "Rethinking Professionalization: Theory, Practice and Professional Identity in Eighteenth-Century German Medicine." *Journal of Modern History* 67 (1995): 835–72.

———. *The Transformation of German Academic Medicine, 1750–1820.* Cambridge: Cambridge University Press, 1996.

Brose, Eric Dorn. *The Politics of Technological Change in Prussia: Out of the Shadow of Antiquity, 1809–1848.* Princeton, NJ: Princeton University Press, 1993.

Brosius, Dieter, "Gerlach Adolf Freiherr v. Münchhausen," in *Neue deutsche Biographie,* 18:523–24.

Bruch, Rüdiger von. "Wissenschaftliche, institutionelle oder politische Innovation? Kameralwissenschaft- Polizeiwissenschaft- Wirthschaftswissenschaft im 18. Jahrhundert im Spiegel der Forschungsgeschichte." In Waszek, *Die Institutionalisierung der Nationalökonomie,* 77–108.

Brückner, Carola, Sylvia Möhle, Ralf Pröve, and Joachim Roschmann. "Vom Fremden zum Bürger: Zuwanderer in Göttingen, 1700–1755." In *Göttingen 1690–1755: Studien zur Sozialgeschichte einer Stadt,* edited by Hermann Wellenreuther, 88–174. Göttingen: Vandenhoeck & Ruprecht, 1988.

Brückner, Jutta. *Staatswissenschaften, Kameralismus und Naturrecht.* Munich: Beck, 1977.

Brunner, Otto, Werner Conze, and Reinhart Koselleck, eds. *Geschichtliche Grundbegriffe.* 8 vols. Stuttgart: Klett, 1975–97.

Budde, Kai. *Wirtschaft, Wissenschaft und Technik im Zeitalter der Aufklarung.* Ubstadt-Weiher: Verlag Regionalkultur, 1993.

Buff, Walter. *Gerlach Adolph Freiherr von Münchhausen als Gründer der Universität Göttingen.* Göttingen: Kaestner, 1937.

Büsching, Anton Friedrich. *Einladungsschrift zu seiner Lehrstunde über die Staatsverfassung der vornehmsten europäischen Reiche.* Halle: Gebauer, 1754.

———. *Erdbeschreibung.* 8th ed. 11 vols. Hamburg: Bohn, 1787.

Carhart, Michael C. *The Science of Culture in Enlightenment Germany.* Cambridge, MA.: Harvard University Press, 2007.

Carlowitz, Hannß Carl von. *Sylvicultura oeconomica: Anweisung zur wilden Baum-Zucht.* 1713. Reprint, Freiberg: TU Bergakademie Freiberg, 2000.

Carlyle, Thomas. *History of Friedrich II of Prussia, Called Frederick the Great.* 8 vols. New York: Lovell, 1890–99.

Carpenter, Kenneth E. *Dialogue in Political Economy: Translations from and into German in the Eighteenth Century.* Boston: Kress Library of Business and Economics, 1977.

Carus, Andre. "Christian Thomasius. Corporatism and the Ethos of the German Professional Classes in the Early Enlightenment." PhD diss., University of Cambridge, 1981.

Catalogus Librorum quos Johannes Beckmannus Professor Gottingens. 4 Februarii 1811 Defunctus. Gottingen: Dieterich, 1812.

[Claproth, Johann Christian]. *Der gegenwärtige Zustand der Göttingischen Universität, in Zween Briefen an einen vornehmen Herrn im Reiche.* Göttingen: Schmidt, 1748.

———. *Schreiben von dem gegenwärtigen Zustande der Göttingischen Universität.* Göttingen: n.p., 1746.

Clark, William. *Academic Charisma and the Origins of the Research University.* Chicago: University of Chicago Press, 2006.

———. "The Death of Metaphysics in Enlightened Prussia." In *The Sciences in Enlightened Europe,* edited by William Clark, Jan Golinski, and Simon Schaffer, 423–73. Chicago: University of Chicago Press, 1999.

———. "On the Ironic Specimen of the Doctor of Philosophy." *Science in Context* 5 (1992): 97–137.

———. "On the Ministerial Archive of Academic Acts." *Science in Context* 9 (1996): 421–86.

Cohen, Claudine. "Leibniz's *Protogaea:* Patronage, Mining, and Evidence for a History of the Earth." In *Proof and Persuasion: Essays on Authorship, Objectivity, and Evidence,* edited by Suzanne Marchand and Elizabeth Lunbeck, 124–43. Turnhout, Belgium: Brepols, 1996.

———. and Andre Wakefield. "Introduction." In Leibniz, *Protogaea.*

Cook, Harold J. *Matters of Exchange: Commerce, Medicine, and Science in the Dutch Golden Age.* New Haven, CT: Yale University Press, 2007.

Cooper, Alix. *Inventing the Indigenous: Local Knowledge and Natural History in Early Modern Europe.* Cambridge: Cambridge University Press, 2007.

———. "'The Possibilities of the Land': The Inventory of 'Natural Riches' in the Early Modern German Territories." *History of Political Economy* 35, supplement (2003): 129–53.

Corran, H. S. *A History of Brewing.* London: David and Charles, 1975.

Dann, Uriel. *Hanover and Great Britain, 1740–1760: Diplomacy and Survival.* Leicester: Leicester University Press, 1991.

Daston, Lorraine. "Strange Facts, Plain Facts, and the Texture of Scientific Experience in the Enlightenment." In *Proof and Persuasion: Essays on Authorship, Ob-*

jectivity, and Evidence, edited by Suzanne Marchand and Elizabeth Lunbeck, 42–59. Turnhout, Belgium: Brepols, 1996.

———. and Peter Galison. *Objectivity.* New York: Zone Books, 2007.

———. and Katherine Park. *Wonders and the Order of Nature, 1150–1750.* New York: Zone Books, 1998.

Dear, Peter, ed. *The Literary Structure of Scientific Argument: Historical Studies.* Philadelphia: University of Pennsylvania Press, 1991.

Debus, Allen G. "Chemists, Physicians, and Changing Perspectives on the Scientific Revolution." *Isis* 89, no. 1 (1998): 66–81.

Delius, Christoph Traugott. *Anleitung zu der Bergbaukunst nach ihrer Theorie und Ausübung, nebst einer Abhandlung von den Grundsätzen der Berg-Kameralwissenschaft.* Vienna: Trattern, 1773.

Deneke, Otto. "Vom alten Grätzel." *Nachrichten von der Grätzel-Gesellschaft zu Göttingen* 2 (1927): 105–10.

Dithmar, Justus Christoph. *Einleitung in die Oeconomische- Policey- und Cameral-Wissenschaften.* Frankfurt an der Oder: Conrad, 1731.

Dittrich, Erhard. *Die deutschen und österreichischen Kameralisten.* Darmstadt: Wissenschaftliche Buchgesellschaft, 1974.

Dopsch, Heinz, Kurt Goldhammer, and Peter Kramml, eds. *Paracelsus (1493–1541): Keines andern Knecht.* Salzburg: Pustet, 1993.

Droysen, Johann Gustav. *Geschichte der Preussischen Politik.* 5 vols. Leipzig: Veit, 1868–86.

Dülfer, Kurt. "Studien zur Organisation des fürstlichen Regierungssystems in der obersten Zentralsphäre im 17. und 18. Jahrhundert." In *Archivar und Historiker, Zum 65. Geburtstag von Heinrich Otto Meisner,* edited by Staatliche Archivverwaltung, 237–53. Berlin: Rütten and Loening, 1956.

Dym, Warren. "Divining Science: Treasure Hunting and the Saxon Mining Industry, 1500–1800." PhD diss., University of California, Davis, 2005.

Ebel, Wilhelm. *Catalogus professorum Gottingensium 1734–1962.* Göttingen: Vandenhoeck & Ruprecht, 1962.

———. ed. *Die Privilegien und Ältesten Statuten der Georg-August-Universität zu Göttingen.* Göttingen: Vandenhoeck & Ruprecht, 1961.

Eck, Reimer. "Göttingen University Library: Excursions into the Archives." *Library History* 9, nos.1–2 (1991): 69–75.

Elias, Norbert. *The Civilizing Process.* New York: Urizen Books, 1978.

Ehrhardt, Siegismund Justus. *Altes und Neues Küstrin.* Glogau: Günther, 1769.

Emmermann, Karl. "Das Zuchthaus zu Celle." PhD diss., Universität Göttingen, 1921.

Engelstein, Laura. "Combined Underdevelopment: Discipline and the Law in Imperial and Soviet Russia." *American Historical Review* 98, no. 2 (1993): 338–53.

Erxleben, Johann Christian Polykarp. *Anfangsgründe der Chemie.* Göttingen: Dieterich, 1775.

———. *Anfangsgründe der Naturgeschichte.* Edited by Johann Friedrich Gmelin. Göttingen: Dieterich, 1782.

Evans, Richard J. *Tales from the German Underworld: Crime and Punishment in the Nineteenth Century.* New Haven, CT: Yale University Press, 1988.

Exner, Wilhelm F. *Johann Beckmann, Begründer der technologischen Wissenschaft.* Vienna: Gerold's Sohn, 1878.

Faller, Gustav. *Gedenkbuch zur hundertjährigen Gründung der königlich-Ungarischen Berg- und Forst-Akademie in Schemnitz, 1770–1870.* Schemnitz: Joerges, 1871.

Fechner, Hermann. "Geschichte des schlesichen Berg- und Hüttenwesens, 1741–1806." *Zeitschrift fur das Berg- Hütten- und Salinenwesens* (1903): 48.

Ferguson, Niall. *The Cash Nexus: Money and Power in the Modern World, 1700–2000.* New York: Basic Books, 2001.

Festschrift zum hundertjährigen Jubiläum der Königl. Sächs. Bergakademie zu Freiberg, am 30 Juli 1866. 2 vols. Dresden: Meinhold, 1866–67.

Festschrift zur 175-Jahrfeier der Bergakademie Clausthal, 1775–1950. Clausthal: Bergakademie Clausthal, 1950.

Feuchtwanger, Lion. *Jud Süss.* Munich: Drei Masken Verlag, 1925.

Finzsch, Norbert, and Robert Jütte, eds. *Institutions of Confinement: Hospitals, Asylums, and Prisons in Western Europe and North America, 1500–1950.* New York: Cambridge University Press, 1996.

Flach, Willy. *Goetheforschung und Verwaltungsgeschichte. Goethe im Geheimen Consilium, 1776–1786.* Weimar: Böhlau, 1952.

Flach, Willy, and Helma Dahl, eds. *Goethe's amtliche Schriften.* 4 vols. Weimar: Böhlau, 1950–87.

Florén, Anders, and Göran Rydén. "The Social Organization of Work at Mines, Furnaces and Forges." In Ågren, *Iron-Making Societies,* 61–138.

Focke, Walther. "Die Lehrmeinungen der Kameralisten über den Handel." PhD diss., Universität Erlangen, 1926.

Forberger, Rudolf. *Die Manufaktur in Sachsen vom Ende des 16. bis zum Ende des 19. Jahrhunderts.* Berlin: Akademie Verlag, 1958.

———. "Zur Rolle und Bedeutung der Bergfabriken in Sachsen." Freiberger Forschungshefte, series D, no. 48, 37–52. Leipzig: Deutscher Verlag für Grundstoffindustrie, 1965.

Foucault, Michel. *Discipline and Punish.* Translated by Alan Sheridan. New York: Vintage Books, 1979.

———. "Governmentality." In *The Foucault Effect,* edited by Graham Burchell, Colin Gordon, and Peter Miller, 87–104. Chicago: University of Chicago Press, 1991.

———. *The Order of Things: An Archaeology of the Human Sciences.* New York: Random House, 1971.

Frängsmyr, Tore, J. L. Heilbron, and Robin E. Rider, eds. *The Quantifying Spirit in the Eighteenth Century.* Berkeley and Los Angeles: University of California Press, 1990.

Frensdorff, Ferdinand. *Die Englischen Prinzen in Göttingen.* Sonderabdruck aus der *Zeitschrift des Historischen Vereins für Niedersachsen.* Göttingen: n.p., 1905.

———. "Georg Brandes, ein hannoverischer Beamter des 18. Jahrhunderts." *Zeitschrift des Historischen Vereins für Niedersachsen* 76 (1911): 1–55.

———. "Über das Leben und die Schriften des Nationalökonomen J. H. G. von Justi." *Nachrichten der Königl. Gesellschaft der Wissenschaften zu Göttingen. Phil.-Hist. Klasse* 4 (1903): 353–503.

———. "Die Vertretung der ökonomischen Wissenschaften in Göttingen vornehmlich im 18. Jahrhundert." In *Festschrift zur Feier des Hundertfünfzigjährigen Bestehens der Königlichen Gesellschaft der Wissenschaften zu Göttingen,* 495–565. Berlin: Wiedmannsche Buchhandlung, 1901.

Freysoldt, August. *Die Fränkischen Wälder im 16. und 17. Jahrhundert.* Steinach: [self-published], 1904.

Friedrich III, Duke of Sachsen-Gotha. *Holz-Taxeordnungen.* Gotha: n.p., 1746.

Frederick II of Prussia. *The Refutation of Machiavelli's Prince, or Anti-Machiavel.* Translated by Paul Sonnino. Athens, OH: Ohio University Press, 1981.

Friedrich, Carl Joachim. "The Continental Tradition of Training Administrators in Law and Jurisprudence." *Journal of Modern History* 11, no. 2 (1939): 129–48.

Fritzsch, Karl-Ewald, and Friedrich Sieber. *Bergmännische Trachten des 18. Jahrhunderts im Erzgebirge und im Mansfeldischen.* Berlin: Akademie Verlag, 1957.

Fröhner, Annette. *Technologie und Enzyklopädismus im Übergang vom 18. zum 19. Jahrhundert: Johann Georg Krünitz (1728–1796) und seine Oeconomisch-technologische Encyklopädie.* Mannheim: J & J Verlag, 1994.

Gagliardo, John. *Reich and Nation: The Holy Roman Empire as Idea and Reality, 1763–1806.* Bloomington: Indiana University Press, 1980.

Gasser, Simon Peter. *Einleitung zu den Oeconomischen, Politischen und Cameral-Wissenschaften.* Halle: Waysenhaus, 1729.

———. *Programma publicum, oder Nöthiger Vorbericht von der von ihro Königl. Maj. in Preussen auf der Universität Halle allergnädigst neu fundirten Profession über die öconomischen, Cameral- und Policey-Wissenschaften.* Halle: n.p., 1728.

Gay, Peter. *The Enlightenment: An Interpretation.* 2 vols. New York: W. W. Norton, 1966–69.

Gedenkbuch zur hundertjährigen Gründung der Königl. Ungarischen Berg- und Forst-Akademie in Schemnitz, 1770–1870. Schemnitz: Joerges, 1871.

Gerhard, Dietrich, ed. *Ständische Vertretungen in Europa im 17. und 18. Jahrhundert.* Göttingen: Vandenhoeck & Ruprecht, 1969.

Gerhard, Hans-Jürgen. "Geld und Geldwert im 18. Jahrhundert." In *Göttingen im 18. Jahrundert,* 25–30. Göttingen: Stadt Göttingen, 1987.

Gierl, Martin. *Geschichte und Organisation: Institutionalisierung als Kommunika-*

tionsprozess am Beispiel der Wissenschaftsakademien um 1900. Göttingen: Vandenhoeck & Ruprecht, 2004.

———. *Pietismus und Aufklärung: Theologische Polemik und die Kommunikationsreform der Wissenschaft am Ende des 17. Jahrhunderts.* Göttingen: Vandenhoeck & Ruprecht, 1997.

———. "Die Universität als Aufklärungsfabrik: Über Kant, gelehrte Ware, Professoren als Fabrikgesellen und darüber, wer die universitätshistorisch herausragende programmatische Schrift des 18. Jahrhunderts in Wirklichkeit geschrieben hat." *Historische Anthropologie* 13, no. 3 (2005): 367–75.

Gleeson, Janet. *The Arcanum: The Extraordinary True Story.* New York: Warner Books, 1998.

Gmelin, Johann Friedrich. *Chemische Grundsätze der Probir- und Schmelzkunst.* Halle: Gebauer, 1786.

———. *Einleitung in die Chemie zum Gebrauch auf Universitäten.* Nürnberg: Raspe, 1780.

———. *Geschichte der Chemie.* 3 vols. Göttingen: n.p., 1797–99.

———. *Grundriß der Mineralogie.* Göttingen: Dieterich, 1790.

———. *Grundsätze der technischen Chemie.* Halle: Gebauer, 1786.

Goldstein, Jan, ed. *Foucault and the Writing of History.* Oxford: Blackwell, 1994.

Gordin, Michael D. "The Importation of Being Earnest: The Early St. Petersburg Academy of Sciences." *Isis* 91, no. 1 (2000): 1–31.

Gorski, Philip S. *The Disciplinary Revolution: Calvinism and the Rise of the State in Early Modern Europe.* Chicago: University of Chicago Press, 2003.

———. "The Protestant Ethic Revisited: Disciplinary Revolution and State Formation in Holland and Prussia." *American Journal of Sociology* 99, no. 2 (1993): 265–316.

Greenfeld, Liah. *The Spirit of Capitalism: Nationalism and Economic Growth.* Cambridge, MA: Harvard University Press, 2001.

Gregory, Brad. "Review: Is Small Beautiful? Microhistory and the History of Everyday Life." *History and Theory* 38, no. 1 (1999): 100–110.

Gross, David. "Temporality and the Modern State." *Theory and Society* 14, no. 1 (1985): 53–82.

Gruber, Johann D., Friedrich C. Neubour, Cyriacus E. Ebell, Christoph H. Pape, Heinrich P. Guden, and Christoph A. Heumann. *Zeit- und Geschicht-Beschreibung der Stadt Göttingen.* 3 vols. Hanover and Göttingen: Förster, 1734–38.

Grünbaum, M. "Drei Hohenzollern-Testamente." *Preußische Jahrbücher* 124 (1906): 61–82.

Guden, Philipp Peter. *Polizey der Industrie, oder Abhandlung von den Mitteln, den Fleiß der Einwohner zu ermuntern, welcher die Königl. Groß-Brittanische Societät der Wissenschaften zu Göttingen i. J. 1766 den Preis zuerkannt hat.* Braunschweig: n.p., 1768.

Gundelach, Ernst. *Die Königlichen Commissarien, Prorectoren und Rektoren der Georg August- Universität zu Göttingen 1734 bis 1957.* Göttingen: Georg-August-Universität, 1957.

Haas, Rudolf. *Die Pfalz am Rhein: 2000 Jahre Landes-, Kultur- und Wirtschaftsgeschichte.* Mannheim: Haas, 1967.

Haasis, Hellmut. G. *Joseph Süß Oppenheimer, genannt Jud Süß: Finanzier, Freidenker, Justizopfer.* Reinbeck: Rowohlt, 1998.

Häberle, Daniel. "Der botanische Garten zu Kaiserslautern." *Pfälzische Geschichtsblätter* 4 (1908): 44–45.

———. "Das 'Landwirtschaftliche Mustergut' zu Siegelbach." *Pfälzische Geschichtsblätter* 6 (1910): 43–46.

Hahn, Roger. *The Anatomy of a Scientific Institution: The Paris Academy of Sciences, 1666–1803.* Berkeley and Los Angeles: University of California Press, 1971.

Hamann, Manfred. "Die alt-hannoverschen Ämter: Ein Überblick." *Niedersächsisches Jahrbuch für Landesgeschichte* 51 (1979): 195–208.

Hamm, Ernst P. "Bureaucratic 'Statistik' or Actualism? K.E.A. von Hoff's 'History' and the History of Geology." *History of Science* 31 (1993): 151–76.

———. "Knowledge from Underground: Leibniz Mines the Enlightenment." *Earth Sciences History* 16 (1997): 77–99.

Hammerstein, Notker. "Die deutschen Universitäten im Zeitalter der Aufklärung." *Zeitschrift für historische Forschung* 10 (1983): 73–89.

———. "Zur Geschichte und Bedeutung der Universitäten im Heiligen Römischen Reich Deutscher Nation." *Historische Zeitschrift* 241 (1985): 287–328.

Happe, Gottlob Christian von. *Der verständige, kluge und ehrliche, auch der unverständige, dumme und betrügerische Saltz- und Berg-Geist oder kurtze doch gründliche Beschreibung des Salzwesens.* N.p., 1717.

———. *Nichts Bessers, als die Accise, Wenn man nur will; Nichts Bösers, als die Accise, Wenn man nicht will; oder Ausführliche Beschreibung, was an der Accise zu loben und zu schelten sey.* Augsburg: n.p., 1717.

Harms, Gerhard. "Editorial." In *Einblicke: Forschungsmagazin der Carl von Ossietzky Universität Oldenburg* 29 (April 1999): 1.

Harnack, Adolf. *Geschichte der Königlich preussischen Akademie der Wissenschaften zu Berlin.* Berlin: Reichsdrückerei, 1900.

Hartung, Fritz, "Goethe als Staatsmann." *Jahrbuch der Goethe-Gesellschaft* 9 (1922): 295–314.

———. *Staatsbildende Kräfte der Neuzeit. Gesammelte Aufsätze.* Berlin: Duncker & Humblot, 1961.

Hausherr, Hans. *Verwaltungseinheit und Ressorttrennung vom Ende des 17. bis zum Beginn des 19. Jahrhunderts.* Berlin: Akademie Verlag, 1955.

Henckel, Johann Friedrich. *Kleine mineralogische und chymische Schrifften.* Edited by Carl Friedrich Zimmermann. Dresden and Leipzig: Hekel, 1744.

Hennings, Klaus Hinrich. "Aspekte der Institutionalisierung der Ökonomie an deutschen Universitäten." In Waszek, *Die Institutionalisierung der Nationalökonomie,* 43–54.

Henschke, Ekkehard. *Landesherrschaft und Bergbauwirtschaft: Zur Wirtschafts- und Verwaltungsgeschichte des Oberharzer Bergbaugebietes im 16. und 17. Jahrhundert.* Berlin: Duncker & Humblot, 1974.

Herrmann, Walther. *Bergbau und Bergleute.* Freiberger Forschungshefte, series D, no. 11. Berlin: Akademie Verlag, 1955.

———. *Bergrat Henckel: Ein Wegbereiter der Bergakademie.* Freiberger Forschungshefte, series D, no. 37. Berlin: Akademie Verlag, 1962.

———. *Goethe und Trebra. Freundschaft und Austausch zwischen Weimar und Freiberg.* Freiberger Forschungshefte, series D, no. 9. Berlin: Akademie Verlag, 1955.

Heß, Heinrich. *Der Thüringer Wald in alten Zeiten. Wald- und Jagdbilder.* Gotha: Perthes, 1898

Heß, Ulrich. *Geheimer Rat und Kabinett in den ernestinischen Staaten Thüringens.* Weimar: Böhlau, 1962.

———. Geschichte der Forstorginisation im Gothaer Land." Unpublished ms., Gotha, 1961.

———."Goethe's amtliche Tätigkeit und ihre dokumentarische Überlieferung." *Archivmitteilungen* 32 (1982): 94–100.

———. ed. *Übersicht über die Bestände des Landesarchivs Gotha.* Weimar: Böhlau, 1960.

Hessenbruch, Arne. "The Spread of Precision Measurement in Scandinavia, 1660–1800." In *The Sciences in the European Periphery During the Enlightenment,* edited by K. Gavroglu, 179–224. Amsterdam: Kluwer, 1999.

Heyne, Christian Gottlob. [no title]. *Göttingische gelehrte Anzeigen* 86 (1810): 849–55.

———. *Rede bei der Trauerfeierlichkeit zur Ehre Seiner Excellenz des Herrn Premierministers Freyherrn von Münchhausen.* Göttingen: n.p., 1770.

[Heynitz, Friedrich Anton von]. *Abhandlung über die Produkte des Mineralreichs in den Königlich-Preußischen Staaten, und über die Mittel diesen Zweig des Staats-Haushaltes immer mehr empor zu bringen.* Berlin: n.p., 1786.

———. "Discours prononcé a l'academie par son Excellence, le Ministre d'etat, Mr. le Baron de Heynitz." *Monats-Schrift der Akademie der Künste und mechanischen Wissenschaften zu Berlin* 1 (January 1788).

———. *Essai d'économie politique.* Basel: Fréres Decker, 1785.

———. *Tabellen über die Staatswirthschaft eines europäischen Staates der vierten größe.* Leipzig: Heinsius, 1786.

Hildebrand, Karl-Gustaf. *Swedish Iron in the Seventeenth and Eighteenth Centuries.* Stockholm: Jernkontoret, 1992.

Hinrichs, Carl. *Preussentum und Pietismus.* Vandenhoeck & Ruprecht, 1971.

Hinrichs, Ernst, ed. *Absolutismus.* Frankfurt am Main: Suhrkamp, 1986.

Hintze, Otto. *Der Beamtenstand.* Leipzig: Teubner, 1971.

———. "Hof- und Landesverwaltung in der Mark Brandenburg unter Joachim II." In *Hohenzollern-Jahrbuch,* 10:138–69. Berlin and Leipzig: Giesecke & Devrient, 1906.

———. *Regierung und Verwaltung.* Edited by Gerhard Oestreich. 3 vols. Göttingen: Vandenhoeck & Ruprecht, 1967.

Hofmann, Johann Andreas. *Unmaßgeblicher Entwurf von dem umfange den Gegenständen, Einrichtungen, Eintheilungen und Verordnungen u. des Polizei-Wesens wie überhaupt im Teutschen-Reiche also auch besonders in den Fürstlichen Hessen-Casselschen Landen mit einer Vorrede von den Polizei-Anstalten in Universitäts-Orten.* Marburg: Weldige, 1765.

Hoffmann, Walter. *Bergakademie Freiberg: Freiberg und sein Bergbau. Die Sächsische Bergakademie.* Frankfurt: Weidlich, 1959.

Holborn, Hajo. *A History of Modern Germany.* 3 vols. New York: Knopf, 1959–69.

Holloran, John. "Professors of Enlightenment at the University of Halle, 1690–1730." PhD diss., University of Virginia, 2000.

Holmes, Frederic L. *Eighteenth-Century Chemistry as an Investigative Enterprise.* Berkeley and Los Angeles: University of California, 1989.

Holquist, Peter. "'Information Is the Alpha and Omega of Our Work': Bolshevik Surveillance in Its Pan-European Context." *Journal of Modern History* 69, no. 3 (1997): 415–50.

Hörnigk, Philipp Wilhelm von. *Oesterreich über alles, wann es nur will.* N.p., 1684.

Höttemann, Walter. "Die Göttinger Tuchindustrie in Vergangenheit und Gegenwart." PhD diss., Universität Göttingen, 1931.

Hubatsch, Walther. *Frederick the Great: Absolutism and Administration.* London: Thames and Hudson, 1975.

Hufbauer, Karl. *The Formation of the German Chemical Community: 1720–1795.* Berkeley and Los Angeles: University of California Press, 1982.

Hull, Isabel V. *Sexuality, State, and Civil Society in Germany, 1700–1815.* Ithaca, NY: Cornell University Press, 1996.

Humpert, Magdalene. *Bibliographie der Kameralwissenschaften.* Köln: Schroeder, 1937.

Ingrao, Charles. *The Hessian Mercenary State: Ideas, Institutions, and Reform under Frederick II, 1760–1785.* Cambridge: Cambridge University Press, 1987.

Jackson, Myles. "Natural and Artificial Budgets: Accounting for Goethe's Economy of Nature." *Science in Context* 7, no. 4 (1994): 409–31.

Jacob, Margaret, and Larry Stewart. *Practical Matter: Newton's Science in the Service of Industry and Empire, 1687–1851.* Cambridge, MA: Harvard University Press, 2004.

Jacobi, Johann Georg. *An die Einwohner der Stadt Zelle.* Halberstadt: n.p., 1770.

Jacobsen, Roswitha. "Die Brüder Seckendorff und ihre Beziehungen zu Sachsen-Gotha." In Jacobsen and Ruge, *Ernst der Fromme,* 95–120.

Jacobsen, Roswitha, and Hans-Jörg Ruge, eds. *Ernst der Fromme (1601–1675): Staatsmann und Reformer.* Bucha bei Jena: Quartus, 2002.

Jakob, Ludwig Heinrich von. *Akademische Freiheit und Disciplin mit besonderer Rücksicht auf die preußischen Universitäten.* Leipzig: n.p., 1819.

———. *Einleitung in das Studium der Staatswissenschaften als Leitfaden für seine Vorlesungen ausgearbeitet vom Staatsrath und Ritter von Jakob Professor in Halle.* Halle: Hemmerde & Schwetschke, 1819.

———. *Grundsätze der Policeygesetzgebung und der Policeyanstalten.* Halle: Ruff, 1809.

———. *Über Cursus und Studien-Plan für angehende Cameralisten.* Halle: Ruff, 1805.

Jars, Gabriel. *Metallurgische Reisen zur Untersuchung und Beobachtung der vornehmsten Eisen- Stahl- Blech- und Steinkohlen-Werke in Deutschland, Schweden, Norwegen, England, und Schottland, vom Jahre 1757 bis 1769.* 4 vols. Translated from the French by Carl Abraham Gerhard. Berlin: Himburg, 1777–85.

Jeserich, Kurt G. A., Hans Pohl, and Georg-Cristoph von Unruh, eds. *Deutsche Vewaltungsgeschichte.* 6 vols. Stuttgart: Deutsche Verlags-Anstalt, 1983–88.

Jeserich, Kurt, Helmut Neuhaus, and Heide Barmeyer-Hartlieb von Wallthor, eds. *Persönlichkeiten der Verwaltung: Biographien zur deutschen Verwaltungsgeschichte, 1648–1945.* Stuttgart: Kohlhammer, 1991.

Johnson, Eric A., and Eric H. Monkkonen, eds. *The Civilization of Crime: Violence in Town and Country Since the Middle Ages.* Champaign-Urbana: University of Illinois Press, 1996.

Johnson, Hubert. "The Concept of Bureaucracy in Cameralism." *Political Science Quarterly* 79, no. 3 (1964): 378–402.

———. *Frederick the Great and His Officials.* New Haven, CT: Yale University Press, 1975.

Jung-Stilling, Johann Heinrich. *The Autobiography of Heinrich Stilling.* Translated by S. Jackson. New York: Harper & Brothers, 1844.

———. *Daß die Kameralwissenschaft auf einer besonders hiezu gestifteten Hohen Schule vorgetragen werden müsse, zum Nuzen der Staaten und der Bürger erörtert.* Lautern: n.p., 1780.

———. *Stilling's Lebensgeschichte.* Vol. 1 of *Johann Heinrich Jungs, genannt Stilling, sämmtliche Werke.* Stuttgart: Scheible, 1841.

———. *Über den Geist der Staatswirthschaft.* Mannheim: n.p., 1787. Reprint, Heidelberg: Manutius, 1990.

———. *Versuch eines Lehrbuchs der Fabrikwissenschaft.* Nürnberg: Grattenauerische Buchhandlung, 1794.

Justi, Johann Heinrich Gottlob von. *Abhandlung über die Frage: Wie die Kupferertze mit Ersparung der Zeit und der Kohlen auf den Kupferhütten besser bearbeitet werden können.* Leipzig: Kummer, 1776.

———. *Abhandlung von denen Manufactur- und Fabriken-Reglements zur Ergän-*

zung seines Werkes von denen Manufacturen und Fabriken. Berlin and Leipzig: Real Schule, 1762.

———. *Abhandlung von den Mitteln die Erkenntniß in den Oeconomischen und Cameral-Wissenschaften dem gemeinen Wesen recht nützlich zu machen.* Göttingen: n.p., 1755.

———. *Auf höchsten Befehl an Sr. Röm. Kaiserl. und zu Ungarn und Böhmen Königl. Majestät erstattetes allerunterthänigstes Gutachten von dem vernünftigen Zusammenhange und practischen Vortrage aller Oeconomischen und Cameralwissenschaften.* Leipzig: n.p., 1754.

———. *Ausführliche Abhandlung von denen Steuern und Abgaben nach ächten, aus dem Endzweck der bürgerlichen Gesellschaften abfliessenden Grundsätzen, und zur Wohlfarth der Völker dienlichen Maaßregeln.* Königsberg and Leipzig: Woltersdorffs Wittwe, 1762.

———. "Erweis, daß das Eisen nicht in dem Eisenerze, oder Steine, vorhanden sey, sondern erst währendem Rösten und Ausschmelzen entstehe." In Justi, *Gesammlete Chymische Schriften*, 1:68–84.

———. *Gesammlete Chymische Schriften, worinnen das Wesen der Metalle und die wichtigsten chymischen Arbeiten vor dem Nahrungsstand und das Bergwesen ausführlich abgehandelt werden.* 2 vols. Berlin and Leipzig: Real Schule, 1760–61.

———. *Gesammelte politische und Finanz-Schriften über wichtige Gegenstände der Staatskunst, der Kriegswissenschaften und des Cameral- und Finanzwesens.* 3 vols. Copenhhagen and Leipzig: n.p., 1761–64; reprint, Aalen: Scientia, 1970.

———. ed. *Göttingische Policey-Amts Nachrichten.* Göttingen, 1755–57 (journal: 3 vols.).

———. *Grundsätze der Policey-Wissenschaft.* Göttingen: Vandenhoeck, 1756.

———. *Leben und Charakter des Königl. Polnischen und Churfürstl. Sächs. Premier-Ministre Grafens von Brühl in verschiedenen Briefen entworfen.* 2 vols. N.p., 1760–61.

———. ed. *Neue Wahrheiten zum Vortheil der Naturkunde und des Gesellschaftlichen Lebens der Menschen.* 2 vols. Leipzig: Breitkopf, 1754–58.

———. *Rechtliche Abhandlung von den Ehen, die an und für sich selbst ungültig und nichtig sind.* Leipzig: Breitkopf, 1757.

———. *Staatswirthschaft, oder systematische Abhandlung aller Oeconomischen und Cameral-Wissenschaften.* 2 vols. Leipzig: Breitkopf, 1755.

———. *System des Finanzwesens, nach vernünftigen aus dem Endzweck der bürgerlichen Gesellschaften und aus der Natur aller Quellen der Einkünfte des Staats hergeleiteten Grundsätzen und Regeln ausführlich abgehandelt.* Halle: Renger, 1766.

———. *Vollständige Abhandlung von den Manufacturen und Fabriken.* 2 vols. Copenhagen: Rothen, 1758–61.

Kaiser, Michael, and Andreas Pecar, eds. *Der zweite Mann im Staat: Oberste*

Amtsträger und Favoriten im Umkreis der Reichsfürsten in der Frühen Neuzeit. Berlin: Duncker & Humblot, 2003.

Kant, Immanuel. *The Conflict of the Faculties. Der Streit der Fakultäten.* 1798. Reprint, translated by Mary J. Gregor, Lincoln: University of Nebraska Press, 1979.

Kästner, Abraham Gotthelf. *Anfangsgründe der angewandten Mathematik.* 2nd ed. Gottingen: Vandenhoeck, 1765.

———. *Anmerkungen über die Markscheidekunst. Nebst einer Abhandlung von Höhenmessungen durch das Barometer.* Göttingen: Vandenhoeck, 1775.

———. *Eine Recension mit Erinnerungen.* N.p, 1780.

———. *Gesinnungen eines deutschen Gelehrten bey der Erinnerung Sr. Excellenz des wohlseel. Premierministers von Münchhausen.* Göttingen: Dieterich, 1770.

Kastner, Sabine. "Wohnen und Bauen in Göttingen." In Wellenreuther, *Studien zur Sozialgeschichte einer Stadt,* 175–251.

Keller, Heinrich. *Tableau von Freyberg.* Frankfurt and Leipzig: n.p., 1786.

Kern, Johann Gottlieb. *Bericht vom Bergbau.* 1769. Edited by Friedrich Wilhelm von Oppel. Reprint, Essen: Verlag Glückauf, 1992.

Kirsten, J. A. G. *Lottchens Reisen ins Zuchthaus.* 3 vols. Leipzig: Müller, 1778.

Kius, Otto. *Das Finanzwesen des Ernestinischen Hauses im sechzehnten Jahrhundert.* Weimar: Böhlau, 1865.

Klammer, Gerhard. "Gewerbeentwicklung und Kulturlandschaft im oberen Leinetal." PhD diss., Göttingen, 1949.

Klein, Ernst. "Johann Heinrich Gottlob von Justi und die preussische Staatswirthschaft." *Vierteljahrschrift für Sozial- und Wirtschaftsgeschichte,* 48 (1961): 145–202.

Klein, Ursula. "The Chemical Workshop Tradition and the Experimental Practice: Discontinuities within Continuities." *Science in Context,* 9, no. 3 (1996): 251–87.

———. *Verbindung und Affinität. Die Grundlegung der neuzeitlichen Chemie an der Wende vom 17. zum 18. Jahrhundert.* Basel: Birkhäuser, 1994.

Klingebiel, Thomas. *Ein Stand für sich? Lokale Amtstraeger in der Fruehen Neuzeit.* Hannover: Hahn, 2002.

Klingenstein, Grete. "Between Mercantilism and Physiocracy." In *State and Society in Early Modern Austria,* edited by Charles W. Ingrao, 181–214. West Lafayette, IN: Purdue University Press, 1994.

Klinger, Andreas. "'Den Staat neu erheben': Zur Staatsbildung Ernsts des Frommen." In Jacobsen and Ruge, *Ernst der Fromme,* 25–34.

———. *Der Gothaer Fürstenstaat. Herrschaft, Konfession und Dynastie unter Herzog Ernst dem Frommen.* Husum: Matthiesen, 2002.

Klinkenborg, Melle. "Die kurfürstliche Kammer und die Begründing des Geheimen Rats in Brandenburg." *Historische Zeitschrift* 114 (1915): 437–88.

Klippel, Diethelm. "Johann August Schlettwein and the Economic Faculty at the University of Gießen." In *La Diffusion Internationale de la Physiocratie,*

XVIIIe-XIXe, edited by Bernard Delmas, Thierry Demals, and Philippe Steiner, 345–65. Grenoble: Presses Universitaires, 1995.

Knemeyer, Franz-Ludwig. "Polizei." *Economy and Society* 9, no. 2 (1980): 172–96.

Koch, Diether. *Das Göttinger Honaratiorentum vom 17. bis zur Mitte des 19. Jahrhunderts.* Göttingen: Vandenhoeck & Ruprecht, 1958.

Koerner, Lisbet. "Daedalus Hyperboreus: Baltic Natural History and Mineralogy in the Enlightenment." In *The Sciences in Enlightened Europe,* edited by William Clark, Jan Golinski, and Simon Schaffer, 389–422. Chicago: University of Chicago Press, 1999.

———. *Linnaeus: Nature and Nation.* Cambridge, MA: Harvard University Press, 1999.

Köhler, Alexander Wilhelm, ed. *Bergmännischer Kalendar.* Freiberg and Annaberg: Craz, 1790.

Köhler, Johann. *Die Keime des Kapitalismus im sächsischen Silberbergbau (1168 bis um 1500).* Freiberger Forschungshefte, series D, no. 13. Berlin: Akademie-Verlag, 1955.

Köhler, Sybilla. "Statistiker in Statistik: Zur Genese der statistischen Disziplin in Deutschland zwischen dem 18. und 20. Jahrhundert." PhD diss., Technische Universität Dresden, 1994.

Kopitzsch, Franklin, ed. *Aufklärung, Absolutismus und Bürgertum in Deutschland.* Munich: Nymphenburger, 1976.

Koschwitz, Hansjürgen. "Die Periodische Wirtschaftspublizistik im Zeitalter des Kameralismus." PhD diss., Universität Göttingen, 1968.

———. "Pressegeschichte einer Universitätsstadt: Entstehung und Aufschwung der periodischen Publizistik Göttingens im frühen 18. Jahrhundert." In *Anfänge Göttinger Sozialwissenschaft,* edited by Hans-Georg Herrlitz and Horst Kern, 150–68. Göttingen: Vandenhoeck & Ruprecht, 1987.

Koselleck, Reinhart. *Preußen zwischen Reform und Revolution.* Stuttgart: Klett, 1967.

Kraemer, Horst. "Der deutsche Kleinstaat des 17. Jahrhunderts im Spiegel von Seckendorffs 'Teutschen Fürstenstaat.'" *Zeitschrift des Vereins für Thüringische Geschichte und Altertumskunde* 33 (1922/24): 1–98.

Kraschewski, Hans-Joachim. "Das Direktionsprinzip im Harzrevier des 17. Jahrhunderts und seine wirtschaftspolitische Bedeutung." In *Vom Bergbau- zum Industrierevier,* edited by Ekkehard Westermann, 125–50. Stuttgart: Steiner, 1995.

———. "Zur Arbeitsverfassung des Goslarer Bergbaus am Rammelsberg im 15. und 16. Jahrhundert." In *Bergbau und Arbeitsrecht: Die Arbeitsverfassung im Europäischen Bergbau des Mittelalters und der frühen Neuzeit,* edited by Karl Heinz Ludwig and Peter Sika, 275–304. Vienna: VWGÖ, 1989.

Krause, Thomas. *Die Strafrechtspflege im Kurfürstentum und Königreich Hannover.* Aalen: Scientia, 1991.

Kriedte, Peter, Hans Medick, and Jürgen Schlumbohm. *Industrialisierung vor der Industrialisierung.* Göttingen: Vandenhoeck & Ruprecht, 1977.

Krünitz, Johann Georg. *Oeconomisch-technologische Encyklopädie oder allgemeines System der Staats- Stadt- Haus- und Landwirthschaft und der Kunstgeschichte in alphabetischer Ordnung.* 128 vols. Berlin: Pauli, 1773–1858.

Lampe, Joachim. *Aristokratie, Hofadel und Staatspatriziat in Kurhannover: Die Lebenskreise der höheren Beamten in den kurhannoverischen Zentral und Hofbehörden.* 2 vols. Göttingen: Vandenhoeck & Ruprecht, 1963.

Lamprecht, Georg Friedrich von. *Über das Studium der Kameralwissenschaften.* Halle: n.p., 1783.

Laudan, Rachel. *From Mineralogy to Geology: The Foundations of a Science, 1650–1830.* Chicago: University of Chicago Press, 1987.

Lauterbach, Werner. *Bergrat Chistlieb Ehregott Gellert.* Freiberger Forschungshefte, series D, no. 200. Leipzig: Deutscher Verlag für Grundstoffindustrie, 1994.

Lehmann, Hartmut. *Das Zeitalter des Absolutismus.* Stuttgart: Kohlhammer, 1980.

Leibniz, G. W. *Protogaea.* Edited and translated by Claudine Cohen and Andre Wakefield. Chicago: University of Chicago Press, 2008.

Lepenies, Wolf. "Wissenschaftsgeschichte und Disziplingeschichte." *Geschichte und Gesellschaft* 4 (1978): 437–51.

Leupold, Jacob. *Theatrum Machinarum Generale.* Leipzig: Zunkel, 1724–25.

Lindemann, Mary. *Health and Healing in Eighteenth-Century Germany.* Baltimore, MD: Johns Hopkins University Press, 1996.

———. *Patriots and Paupers: Hamburg, 1712–1830.* New York: Oxford University Press, 1990.

Lindenfeld, David F. *The Practical Imagination: The German Sciences of State in the Nineteenth Century.* Chicago: University of Chicago Press, 1997.

Linke, Uli. "Folklore, Anthropology, and the Government of Social Life." *Comparative Studies in Society and History* 32, no. 1 (1990): 117–48.

Linnaeus, Carl von. *Illustris Caroli a Linné terminologia conchyliologiae.* Edited by Johann Beckmann. Göttingen: Vandenhoeck, 1772.

Losch, Phillip. "Die hessischen Prinzen in Göttingen, 1754–56." *Nachrichten von der Grätzel-Gesellschaft zu Göttingen,* 2 (1927): 28–34.

Lotz, Albert. *Geschichte des Deutschen Bürgertums.* Berlin: Decker, 1909.

Lowood, Henry E. "The Calculating Forester: Quantification, Cameral Science, and the Emergence of Scientific Forestry Management in Germany." In Frängsmyr, Heilbron, and Rider, *Quantifying Spirit,* 315–42.

———. *Patriotism, Profit, and the Promotion of Science in the German Enlightenment.* New York: Garland, 1991.

Ludewig, Johann Peter von. *Die, von Sr. Königlichen Majestät, unserm allergnädigsten Könige, auf dero Universität Halle, am 14 Julii 1727 neu angerichtete Profession in Oeconomie, Policey, und Cammer-Sachen.* Halle: Neue Buchhandlung, 1727.

Lüdtke, Alf. *"Gemeinwohl," Polizei und "Festungspraxis": Staatliche Gewaltsamkeit und innere Verwaltung in Preussen, 1815–1850.* Göttingen: Vandenhoeck & Ruprecht, 1982.

Luh, Jürgen. "Vom Pagen zum Premierminister: Graf Heinrich von Brühl (1700–1763) und die Gunst der sächsisch-polnischen Kurfürsten und Könige August II. und August III." In Kaiser and Pecar, *Der zweite Mann im Staat,* 121–35.

Lünig, Johann Christian, ed. *Codex Augusteus oder Neuvermehrtes Corpus Juris Saxonici.* Leipzig: Gleditsch, 1724.

MacDonald, Scott B., and Albert L Gastmann. *A History of Credit and Power in the Western World.* New Brunswick, NJ: Transaction, 2001.

[Machiavel, Maria]. *Der volkommene Kameraliste: Entworfen von Maria Machiavel aus der Italienischen Urschrift des Verfassers ins Teutsche übersetzt von U.* Frankfurt and Leipzig: n.p., 1764.

Mahoney, Susan K. "A Good Constitution: Social Science in Eighteenth-Century Göttingen." PhD diss., University of Chicago, 1982.

Maier, Hans. *Die ältere Deutsche Staats- und Verwaltungslehre.* 2nd ed. Munich: Beck, 1980.

Mann, Fritz Karl. *Steuerpolitische Ideale: Vergleichende Studien zur Geschichte der ökonomischen und politischen Ideen und ihres wirkens in der öffentlichen Meinung, 1600–1935.* Stuttgart: Fischer, 1978.

Marino, Luigi. *Praeceptores Germaniae: Göttingen, 1770–1820.* Göttingen: Vandenhoeck & Ruprecht, 1995.

Martin, Guntram. "Bergverfassung, Bergverwaltung, Bergrecht im sächsischen Montanwesen des 19. Jahrhunderts: Probleme des Überganges vom Direktionsprinzip zur freien Unternehmerwirthschaft (1831 bis 1868)." PhD diss., Technische Universität Dresden, 1994.

Marx, Karl, and Friedrich Engels. *Karl Marx, Friedrich Engels: Werke.* 39 vols. Berlin: Dietz, 1961–74.

McClelland, Charles E. *State, Society, and University in Germany, 1700–1914.* Cambridge: Cambridge University Press, 1980.

McNeely, Ian F. *The Emancipation of Writing: German Civil Society in the Making, 1790s–1820s.* Berkeley and Los Angeles: University of California Press, 2003.

Medick, Hans. *Weben und Überleben in Laichingen, 1650–1900.* Göttingen: Vandenhoeck & Ruprecht, 1996.

Medicus, Friedrich Casimir. *Beiträge zur schönen Gartenkunst.* Mannheim: Hof- und akademische Buchhandlung, 1782.

———. *Kleine ökonomische Aufsätze.* Mannheim: Schwan & Götz, 1804.

———. *Nachricht an das Publikum.* Mannheim: Hof- und akademische Buchhandlung, 1784.

———. *Philosophische Botanik, mit kritischen Bemerkungen.* Mannheim: Hof- und akademische Buchhandlung, 1789.

———. *Von dem Bau auf Steinkohlen.* Mannheim: n.p., 1768.

Meier, Ernst von. *Hannoversche Verfassungs- und Verwaltungsgeschichte, 1680–1866.* Leipzig: Duncker & Humblot, 1898–99.

Meinel, Christoph. "Reine und angewandte Chemie." *Berichte zur Wissenschaftsgeschichte* 8 (1985): 25–45.

Meiners, Christoph. *Geschichte der Entstehung und Entwicklung der hohen Schulen unsers Erdtheils.* 4 vols. Göttingen: Röwer, 1802–5.

———. *Über die Verfassung und Verwaltung deutscher Universitäten.* 2 vols. Göttingen: Röwer, 1801–2.

Melzheimer, Werner. *Die Festung und Garnison Küstrin.* Berlin: Möller, 1989.

[Michaelis, Johann David]. *Raisonnement über die protestantischen Universitäten in Deutschland.* 4 vols. Frankfurt: n.p., 1768–76.

Miller, Peter N. "Nazis and Neo-Stoics: Otto Brunner and Gerhard Oestreich before and after the Second World War." *Past and Present* 176, no. 1 (2002): 144–86.

Minchinton, Walter E. *The British Tinplate Industry: A History.* London: Oxford University Press, 1957.

———. "The Diffusion of Tinplate Manufacture." *Economic History Review,* n.s., 9, no. 2 (1956): 349–58.

Mittenzwei, Ingrid. *Preussen nach dem Siebenjährigen Krieg.* Berlin: Akademie Verlag, 1979.

Mohnhaupt, Heinz. *Die Göttinger Ratsverfassung vom 16. bis 19. Jahrhundert.* Göttingen: Vandenhoeck & Ruprecht, 1965.

Mokyr, Joel. *The Gifts of Athena: Historical Origins of the Knowledge Economy.* Princeton, NJ: Princeton University Press, 2002.

———. "The Intellectual Origins of Modern Economic Growth." *Journal of Economic History* 65, no. 2 (2005): 285–351.

Möller, Horst. *Fürstenstaat oder Bürgernation: Deutschland 1763–1815.* Berlin: Siedler, 1989.

Moshammer, Franz Xaver. *Gedanken und Vorschläge über die neuesten Anstalten teutscher Fürsten die Kameralwissenschaften auf hohen Schulen in Flor zu bringen.* Regensburg: Montag, 1782.

Müller, Emil. *Zur Geschichte des höheren Schulwesens.* Kaiserslautern: Crusius, 1899.

Multhauf, Robert P. 1966. *The Origins of Chemistry.* New York: Watts.

Münch, Paul. *Das Jahhundert des Zwiespalts: Deutsche Geschichte, 1600–1700.* Stuttgart: Kohlhammer, 1999.

Naudé, Wilhelm. "Zur Geschichte des preußischen Subalternbeamtentums." *Forschungen zur Brandenburgischen und Preußischen Geschichte* 18 (1905): 365–86.

Napp-Zinn, Anton Felix. *Johann Friedrich Pfeiffer und die Kameralwissenschaften an der Universität Mainz.* Wiesbaden: Steiner, 1955.

Naumann, Friedrich, ed. *Georgius Agricola—500 Jahre.* Basel: Birkhäuser, 1994.

Naumann, Viktor. *Die deutschen Universitäten in ihrem Verhältnis zum Staat, ihre Verfassung und Verwaltung, ihre Statuten und Disziplinar-Ordnungen.* Graz: Styria, 1909.

Nef, John U. "Mining and Metallurgy in Medieval Civilization." In *The Cambridge Economic History of Europe,* edited by M. M. Postan and Edward Miller, 2:693–756. Cambridge: Cambridge University Press, 1952.

———. "Silver Production in Central Europe, 1450–1618." *Journal of Political Economy* 49 (1941):575–91.

Neue deutsche Biographie. 19 vols. Berlin: Duncker & Humblot, 1953–99.

Neugebauer, Wolfgang. *Politischer Wandel im Osten: Ost- und Westpreussen von den alten Ständen zum Konstitutionalismus.* Stuttgart: Steiner, 1992.

Nielsen, Axel. *Die Entstehung der deutschen Kameralwissenschaft im 17. Jahrhundert.* Jena: Fischer, 1911.

Nissen, Walter. *Göttingen Gestern und Heute: Eine Sammlung von Zeugnissen zur Stadt- und Universitätsgeschichte.* Göttingen: Stadt Göttingen, 1972.

Novick, Peter. *That Noble Dream: The "Objectivity Question" and the American Historical Profession.* Cambridge: Cambridge University Press 1988.

Nummedal, Tara. *Alchemy and Authority in the Holy Roman Empire.* Chicago: University of Chicago Press, 2007.

Oberschelp, Reinhard. *Niedersachsen, 1760–1820: Wirthschaft, Gesellschaft, Kultur im Land Hannover und Nachbargebieten.* Hildesheim: Lax, 1982.

Oestreich, Gerhard. *Neostoicism and the Early Modern State.* Edited by Brigitta Oestreich and H. G. Koenigsberger. Translated by David McLintock. Cambridge: Cambridge University Press, 1982.

———. "Das persönliche Regiment der deutschen Fürsten am Beginn der Neuzeit." In *Die Welt als Geschichte* 1 (1935): 218–37, 300–316.

———. *Strukturprobleme der frühen Neuzeit. Ausgewählte Aufsätze.* Edited by Brigitta Oestreich. Berlin: Duncker & Humblot, 1980.

Ogilvie, Sheilagh C., ed. *Germany: A New Social and Economic History.* 2 vols. London: Arnold, 1996–99.

———. "Germany and the Seventeenth-Century Crisis." *Historical Journal* 35, no. 2 (1992): 417–41.

———. "The State in Germany: A Non-Prussian View." In Brewer and Hellmuth, *Rethinking Leviathan*, 167–202.

Ogilvie, Sheilagh, and Markus Cerman, eds. *European Proto-Industrialization.* Cambridge: Cambridge University Press, 1996.

Ohnsorge, Werner. "Zum Problem: Fürst und Verwaltung um die Wende des 16. Jahrhunderts." *Blätter für deutsche Landesgeschichte* 88 (1951): 150–74.

———. "Zur Entstehung und Geschichte der Geheimen Kammerkanzlei im albertinischen Kursachsen." *Neues Archiv für Sächsiche Geschichte* 61 (1940): 158–215.

Ospovat, Alexander. "The Importance of Regional Geology in the Geological Theories of Abraham Gottlob Werner: A Contrary Opinion." *Annals of Science* 37, no. 4 (1980): 433–40.

———. "Romanticism and Geology: Five Students of Abraham Gottlob Werner." *Eighteenth-Century Life* 7 (1982):105–17.

Osterloh, Karl-Heinz. *Joseph von Sonnenfels und die österreichische Reformbewegung im Zeitalter des aufgeklärten Absolutismus.* Lübeck: Matthiesen, 1970.

Outram, Dorinda. *The Enlightenment.* Cambridge: Cambridge University Press, 1995.

Oz-Salzberger, Fania. *Translating the Enlightenment: Scottish Civic Discourse in Eighteenth-Century Germany.* Oxford: Oxford University Press, 1995.

Pahner, Richard. "Veit Ludwig von Seckendorff und seine Gedanken über Erziehung und Unterricht." PhD diss., University of Leipzig, 1892.

Papperitz, Erwin. *Gedenkschrift zum hundertfünfzigjährigen Jubiläum der Königlich Sächsischen Bergakademie zu Freiberg.* Freiberg: Gerlach, 1916.

Partington, James R. *A History of Chemistry.* 4 vols. London: Macmillan, 1961–1970.

Paulsen, Friedrich. *Das deutsche Bildungswesen in seiner geschichtlichen Entwicklung.* Leipzig: Teubner, 1906.

———. *Geschichte des Gelehrten Unterrichts auf den deutschen Schulen und Universitäten.* 2 vols. 2nd ed. Leipzig: Veit, 1896.

Plettenberg, Alexandra. "Die Hohe-Kameral-Schule zu Lautern, 1774–1784." PhD diss., Ludwig Maximilian University, Munich, 1983.

Pockels, Carl Friedrich, ed. *Denkwürdigkeiten zur Bereicherung der Erfahrungsseelenlehre und Charakterkunde: Ein Lesebuch für Gelehrte und Ungelehrte.* Halle: Renger, 1794.

Poller, Oskar. *Schicksal der ersten Kaiserslauterer Hochschule und ihrer Studierenden.* Ludwigshafen: Arbeitsgemeinschaft Pfälzisch-Rheinische Familienkunde, 1979.

Porter, Theodore M. *Trust in Numbers: The Pursuit of Objectivity in Science and Public Life.* Princeton, NJ: Princeton University Press, 1995.

Postan, M. M., and E. E. Rich, eds. *Trade and Industry in the Middle Ages.* Vol. 2 of *The Cambridge Economic History of Europe.* Cambridge: Cambridge University Press, 1952.

Pott, Johann Heinrich. *Animadversiones physico-chimicae circa varias hypotheses et experimenta Elleri.* Berlin: "At cost of the author," 1756.

———. *Chymische Untersuchungen welche fürnehmlich von der Lithogeognosia handeln.* Berlin: Voss, 1757.

———. *Send-Schreiben an den Herrn Berg-Rath von Justi.* Berlin: n.p., 1760.

Prescher, Hans, and Otfried Wagenbreth. *Georgius Agricola: Seine Zeit und ihre Spuren.* Leipzig: Deutscher Verlag für Grundstoffindustrie, 1994.

Pütter, Johann Stephan, Friedrich Saalfeld, and Georg Heinrich Oesterley. *Versuch einer academischen Gelehrtengeschichte von der Georg-Augustus-Universität zu Göttingen.* 4 vols. Göttingen: Vandenhoeck, 1765–1838.

Raeff, Marc. *The Well-Ordered Police State: Social and Institutional Change through Law in the Germanies and Russia, 1600–1800.* New Haven, CT: Yale University Press, 1983.

———. "The Well-Ordered Police State and the Development of Modernity in

Seventeenth- and Eighteenth-Century Europe: An Attempt at a Comparative Approach." *American Historical Review* 80, no. 5 (1975): 1221–43.

Rall, Hans. *Kurfürst Karl Theodor.* Mannheim: Wissenschaftsverlag, 1993.

Ramati, Ayval. "Harmony at a Distance: Leibniz's Scientific Academies." *Isis* 87, no. 3 (1996): 430–52.

Rau, Karl Heinrich. *Ueber die Kameralwissenschaft. Entwicklung ihres Wesens und ihrer Theile.* Heidelberg: C. F. Winter, 1823.

Raupach, Angela. "Zum Verhältnis von Politik und Ökonomie im Kameralismus: Ein Beitrag zur sozialen Theoriebildung in Deutschland in ihrer Genese als Polizei." PhD diss., Universität Hamburg, 1982.

Recktenwald, Horst Claus. "Cameralism." In *The New Palgrave: A Dictionary of Economics,* edited by John Eatwell, Murray Milgate, Peter Newman, and Robert H. I. Palgrave, 1:313–14. London: Macmillan, 1987.

Reinwald, Jochen. "Die Ursachen der Beseitigung des Direktionsprinzips im sächsischen Silberbergbau." PhD diss., Technische Universität Freiberg, 1967.

Remer, Justus. *Johann Heinrich Gottlob Justi: Ein deutscher Volkswirt des 18. Jahrhunderts.* Stuttgart: Kohlhammer, 1938.

Ringer, Fritz K. *The Decline of the German Mandarins: The German Academic Community, 1890–1933.* Cambridge, MA: Harvard University Press, 1969.

Riotte, Torsten. "George III and Hanover." In Simms and Riotte, *The Hanoverian Dimension in British History, 1714–1837,* 58-85.

Roberts, Lissa. "Filling the Space of Possibilities: Eighteenth-Century Chemistry's Transition from Art to Science." *Science in Context* 6, no. 2 (1993): 511–53.

———. "Setting the Table: The History of Eighteenth-Century Chemistry as Read Through Its Tables." In Dear, *Literary Structure,* 99–132.

Roscher, Wilhelm. *Geschichte der Nationalökonomik in Deutschland.* Munich: Oldenburg, 1874.

Rosenband, Leonard N. "Becoming Competitive: England's Papermaking Apprenticeship." In *The Mindful Hand: Inquiry and Invention from the Late Renaissance to Early Industrialisation,* edited by Lissa Roberts, Simon Schaffer, and Peter Dear, 378–402. Amsterdam: Royal Netherlands Academy of Sciences, 2007.

———. "Never Just Business: David Landes, *The Unbound Prometheus.*" *Technology and Culture* 46, no. 1 (2005): 168–76.

———. *Papermaking in Eighteenth-Century France: Management, Labor, and Revolution at the Montgolfier Mill, 1761–1805.* Baltimore, MD: Johns Hopkins University Press, 2000.

Rosenberg, Hans. *Bureaucracy, Aristocracy, and Autocracy. The Prussian Experience, 1660–1815.* Cambridge, MA: Harvard University Press, 1958.

Rosenthal, Eduard. "Die Behördenorganisation Kaiser Ferdinands I., das Vorbild der Verwaltungsorganisation in den deutschen Territorien." *Archiv für österreichische Geschichte* 69, no. 1 (1887): 51–316.

Rössler, Emil, ed. *Die Gründung der Universität Göttingen.* Göttingen: Vandenhoeck & Ruprecht, 1855.

Rudwick, Martin J. S. *Bursting the Limits of Time: The Reconstruction of Geohistory in the Age of Revolution.* Chicago: University of Chicago Press, 2005.

Ruge, Hans-Jörg. "Übersicht über die Besoldung." In Jacobsen and Ruge, *Ernst der Fromme,* 121–26.

———. "Vom Bibliothekar zum Geheimen Rat: Aspekte der beruflichen Laufbahn Veit Ludwig von Seckendorffs (1626–1692) in den Jahren seiner Anstellung im sachsen-gothaischen Staatsdienst (1646–1664)." Unpublished ms. Abschlußarbeit, Archivwissenschaft, Leipzig, 1992.

Rulffs, August Friedrich. *Ueber die Preisfrage der Königl. Societät der Wissenschaften zu Göttingen von der vortheilhaftesten Einrichtung der Werk- und Zuchthäuser.* Göttingen: Rosenbusch, 1783.

Saathoff, Albrecht. *Geschichte der Stadt Göttingen.* 2 vols. Göttingen: Vandenhoeck & Ruprecht, 1937.

Sabean, David. *Kinship in Neckarhausen, 1700–1870.* Cambridge: Cambridge University Press, 1998.

Sachse, Wieland. *Göttingen im 18. und 19. Jahrhundert: Zur Bevölkerungs- und Sozialstruktur einer deutschen Universitätsstadt.* Göttingen: Vandenhoeck & Ruprecht, 1987.

Sammlung der Verordnungen für das Königreich Hannover aus der Zeit vor dem Jahre 1813. Edited by Christian Hermann Ebhardt. 3 vols. Hannover: Telgenersche Hofbuchdruckerei, 1855.

Sandl, Marcus. *Ökonomie des Raumes: Der kameralwissenschaftliche Entwurf der Staatswirtschaft im 18. Jahrhundert.* Cologne: Böhlau, 1999.

Sauter, Michael J. "Clockwatchers and Stargazers: Time Discipline in Early Modern Berlin." *American Historical Review* 112, no. 3 (2007): 685–709.

Schabas, Margaret, and Neil De Marchi. "Introduction to *Oeconomies in the Age of Newton.*" *History of Political Economy* 35 (2003, annual supplement): 1–13.

Schama, Simon. *The Embarrassment of Riches: An Interpretation of Dutch Culture in the Golden Age.* New York: Knopf, 1987.

Schellhas, Walter. "Abraham Gottlob Werner als Inspektor der Bergakademie Freiberg und als Mitglied des Sächsischen Oberbergamtes zu Freiberg." Freiberger Forschungshefte, Series C, no. 22, 245–78. Leipzig: Deutscher Verlag für Grundstoffindustrie, 1967.

———. "Eine 'Bergwercks-Academie' in Bräunsdorf bei Freiberg/Sa.?" Freiberger Forschungshefte, series D, no. 22, 57–185. Leipzig: Deutscher Verlag für Grundstoffindustrie, 1957.

Schelsky, Helmut. *Einsamkeit und Freiheit: Idee und Gestalt der deutschen Universität und ihrer Reformen.* Reinbeck bei Hamburg: Rowohlt, 1963.

Schikorsky, Isa. "Das Collegium Carolinum als Reformanstalt: Der beschwerliche Weg zwischen Lateinschule und Universität." In *Technische Universität Braun-*

schweig: Vom Collegium Carolinum zur Technischen Universität, 1745–1995, edited by Peter Albrecht and Walter Kertz. Hildesheim: Olms, 1995.

Schilling, Heinz. *Höfe und Allianzen: Deutschland, 1648–1763.* Berlin; Siedler, 1989.

———. *Kirchenzucht und Sozialdisziplinierung im frühneuzeitlichen Europa.* Berlin : Duncker & Humblot, 1994.

Schlechte, Horst, ed. *Die Staatsreform in Kursachsen, 1762–1763: Quellen zum Kursächsischen Retablissement nach dem Siebenjährigen Kriege.* Berlin: Rütten & Loening, 1958.

Schmidt-Bielicke, Hans. "Der Autarkiegedanke im Merkantilismus." PhD diss., Universität Jena, 1933.

Schmoller, Gustav von, ed. *Die Behördenorganisation und die allgemeine Staatsverwaltung Preussens im 18. Jahrhundert.* 16 vols. Berlin: Parey, 1894–1970.

———. *The Mercantile System and Its Historical Significance.* Translated by W. J. Ashley. New York: Macmillan, 1897.

———. "Der preußische Beamtenstand unter Friedrich Wilhelm I." *Preußische Jahrbücher,* 26 (1870): 148–72, 253–70, 538–55.

———. Otto Krauske, Victor Loewe, Wilhelm Stolze, Otto Hintze, Martin Friedrich, Wilhelm Hass, Wolfgang Peters, Ernst Posner, Peter Baumgart, and Gerd Heinrich, eds. *Acta Borussica: Die Behördenorganisation und die allgemeine Staatsverwaltung Preussens im 18. Jahrhundert.* 16 vols. Berlin: Parey, 1894–1982.

Schönberg, Abraham von. *Ausführliche Berg-Information.* Leipzig and Zwickau: n.p., 1693.

Schnabel, Franz. *Abhandlungen und Vorträge, 1914–1965.* Freiburg: Herder, 1970.

Schrader, Wilhelm. *Geschichte der Friedrichs-Universität zu Halle.* 2 vols. Berlin: Dummler, 1894.

Schreber, Daniel G., ed. *Neue Sammlung verschiedener in die Cameralwissenschaften einschlagender Abhandlungen und Urkunden, auch andrer Nachrichten.* 8 vols. Bützow and Wismar: Berger and Boedner, 1762–65.

———. *Sammlung verschiedener Schriften, welche in die öconomischen, Policey- und Cameral- auch andere Wissenschaften einschlagen.* 16 vols. Halle: Curts, 1755–65.

———. *Zwo Schriften von der Geschichte und Nothwendigkeit der Cameralwissenschaften in so ferne sie als Universitätswissenschaften anzusehen sind.* Leipzig: Dyck, 1764.

Schröder, Wilhelm Freiherr von. *Fürstliche Schatz- und Rentkammer: nebst seinem Tractat von Goldmachen.* 1752. Reprint, Vaduz/Liechtenstein: Topos, 1978.

Schultze, Walther. *Geschichte der preussischen Regieverwaltung von 1766 bis 1786: Ein historisch-kritischer Versuch.* Leipzig: Duncker & Humblot, 1888.

Schumpeter, Joseph. "The Crisis of the Tax State." 1918. In *International Economic Papers,* edited by Alan Peacock, Wolfgang Stolper, Ralph Turvey, and Elizabeth Henderson, 4: 5–38. New York: Macmillan, 1954.

———. *History of Economic Analysis.* Edited by Elizabeth B. Schumpeter. New York: Oxford University Press, 1954.

Scott, James C. *Seeing Like a State: How Certain Schemes to Improve the Human Condition Have Failed.* New Haven, CT: Yale University Press, 1998.

Seckendorff, Veit Ludwig von. *Lob-Rede des Heunßel-Bergs.* Gotha: n.p., 1702.

———. *Teutscher Fürsten Stat.* 1665. 2 vols. Reprint, Glashütten: Auvermann, 1976.

Seelig, Eckhard. *Die Entstehung des Direktionsprinzips im Sächsischen Bergrecht und seine Weiterentwicklung im Merkantilismus.* Göttingen: Funke, 1971.

Selle, Götz von. *Die Georg-August Universität zu Göttingen, 1737–1937.* Göttingen: Vandenhoeck & Ruprecht, 1937.

Shapin, Steven, and Simon Schaffer. *Leviathan and the Air-Pump: Hobbes, Boyle, and the Experimental Life.* Princeton, NJ: Princeton University Press, 1985.

Sheehan, James J. *German History, 1770–1866.* Oxford: Clarendon Press, 1989.

Shovlin, John. *The Political Economy of Virtue: Luxury, Patriotism, and the Origins of the French Revolution.* Ithaca, NY: Cornell University Press, 2006.

Simms, Brendan, and Torsten Riotte, eds. *The Hanoverian Dimension in British History, 1714–1837.* Cambridge: Cambridge University Press, 2007.

Small, Albion. *The Cameralists: The Pioneers of German Social Polity.* Chicago: University of Chicago Press, 1909.

———. "What Is a Sociologist?" *American Journal of Sociology* 8, no. 4 (1903): 468–77.

Smith, Adam. *An Inquiry into the Nature and Causes of the Wealth of Nations.* 1776. Edited by Edwin Cannan. Chicago: University of Chicago Press, 1976.

———. *The Theory of Moral Sentiments.* 1759. Edited by D. D. Raphael and A. L. Macfie. Oxford: Oxford University Press, 1976.

Smith, Cyril Stanley. "The Discovery of Carbon in Steel." *Technology and Culture* 5, no. 2 (1964): 149–75.

Smith, Pamela H. "Alchemy as a Language of Mediation at the Habsburg Court." *Isis* 85, no. 1 (1994): 1–25.

———. *The Business of Alchemy: Science and Culture in the Holy Roman Empire.* Princeton, NJ: Princeton University Press, 1994.

Soetbeer, Adolf. *Edelmetall-Produktion und Wertverhältniss zwischen Gold und Silber seit der Entdeckung Amerika's bis zur Gegenwart.* Gotha: Perthes, 1879.

Sommer, Louise. *Die österreichischen Kameralisten in dogmengeschichtlicher Darstellung.* 2 vols. Vienna: Konegen, 1920–25.

Sperber, Jonathan. "State and Society in Prussia: Thoughts on a New Edition of Reinhart Koselleck's *Preussen zwischen Reform und Revolution.*" *Journal of Modern History* 57, no. 2 (1985): 278–96.

Spierenburg, Pieter. *The Prison Experience: Disciplinary Institutions and Their Inmates in Early Modern Europe.* New Brunswick, NJ: Rutgers University Press, 1991.

Springer, Johann Cristoph Erich. "Review of *Der volkommene Kameraliste.*" *Allgemeine deutsche Bibliothek* 10, no. 1 (1769): 296–97.

———. *Über die Protestantischen Universitäten in Deutschland.* Strasburg: n.p., 1769.

Stahlschmidt, Jens Wilhelm. "Policey und Fürstenstaat: Die gothaische Policeygesetzgebung unter Herzog Ernst dem Frommen." PhD diss., Ruhr-Universität Bochum, 1999.

Steenbuck, Kurt. *Silber und Kupfer aus Ilmenau: Ein Bergwerk unter Goethe's Leitung.* Weimar: Böhlau, 1995.

Stichweh, Rudolf. "The Sociology of Scientific Disciplines: On the Genesis and Stability of the Disciplinary Structure of Modern Science." *Science in Context* 5, no. 1 (1992): 3–15.

———. *Zur Entstehung des modernen Systems wissenschaftlicher Disziplinen: Physik in Deutschland, 1740–1890.* Frankfurt: Suhrkamp, 1984.

Stieda, Wilhelm. *Die Nationalökonomie als Universitätswissenschaft.* Leipzig: Teubner, 1906.

———. "Zur Geschichte der Hohen Kameralschule in Kaiserslautern." *Zeitschrift fur die Geschichte des Oberrheins* 25 (1910): 340–53.

Strunz, Hugo. *Von der Bergakademie zur Technischen Universität Berlin, 1770–1970.* Berlin: Technische Universität Berlin, 1970.

Sturmhoefel, Konrad. *Illustrierte Geschichte des Albertinischen Sachsen.* 3 vols. Leipzig: Hubel & Denk, 1909.

Szabo, Franz. *Kaunitz and Enlightened Absolutism, 1753–1780.* Cambridge: Cambridge University Press, 1994.

Szabo, Franz, and Grete Klingenstein, eds. *Staatskanzler Wenzel Anton von Kaunitz-Rietberg: 1711–1794.* Graz: Schnider, 1996.

Tautscher, Anton. *Geschichte der deutschen Finanzwissenschaft.* Tübingen: n.p., 1952.

Torinus, Heinz. *Die Entstehung des preussichen Beamtenethos unter Friedrich Wilhelm I: Ein Beitrag zur Geschichte der Staatserziehung.* Würzburg: Mayr, 1935.

Trebra, Friedrich Wilhelm Heinrich von. *Bergmeister Leben und Wirken in Marienberg.* 1818. Reprint, Leipzig: VEB Deutscher Verlag, 1990.

———. *Erklärungen der Bergwerks-Charte von dem wichtigsten Theil der Gebürge im Bergamtsrevier Marienberg.* Annaberg: n.p.,1770.

Tribe, Keith. "Cameralism and the Science of Government." *Journal of Modern History* 56, no. 2 (1984): 263–84.

———. *Governing Economy: The Reformation of German Economic Discourse, 1750–1840.* Cambridge: Cambridge University Press, 1988.

———. "Die 'Kameral Hohe Schule zu Lautern' und die Anfänge der ökonomischen Lehre in Heidelberg (1744–1822)." In Waszek, *Die Institutionalisierung der Nationalökonomie,* 162–91.

Troitzsch, Ulrich. *Ansätze technologischen Denkens bei den Kameralisten des 17. und 18. Jahrhunderts.* Berlin: Duncker & Humblot, 1966.

Turner, R. Steven. "The Bildungsbürgertum and the Learned Professions in Prussia, 1770–1830: The Origins of a Class." *Histoire Social/Social History* 13 (1980):105–35.

———. "The Great Transition and the Social Patterns of German Science." *Minerva* 25, nos. 1–2 (1987): 56–76.

———. "The Prussian Universities and the Research Imperative, 1806 to 1848." PhD diss., Princeton University, 1973.

———. "University Reformers and Professorial Scholarship in Germany, 1760–1806." In *The University in Society,* edited by Lawrence Stone, 2: 495–531. 2 vols. Princeton, NJ: Princeton University Press, 1974.

Tweedale, Geoffrey. "Metallurgy and Technological Change: A Case Study of Sheffield Specialty Steel and America, 1830–1930." *Technology and Culture* 27, no. 2 (1986): 189–222.

Vann, James A. *The Making of a State: Würtemberg, 1593–1793.* Ithaca, NY: Cornell University Press, 1984.

———. "New Directions for Study of the Old Reich." *Journal of Modern History* 58, supplement (1986): S3–S22.

———. *The Swabian Kreis: Institutional Growth in the Holy Roman Empire, 1648–1715.* Brussels: Librairie Encyclopedique, 1975.

Vierhaus, Rudolf. *Germany in the Age of Absolutism.* Translated by Jonathan B. Knudsen. Cambridge: Cambridge University Press, 1988.

———. *Staaten und Stände: Vom Westfälischen Frieden bis zum Hubertusburger Frieden, 1648–1763.* Berlin: Propyläen, 1984.

Voigt, Julius. *Goethe und Ilmenau.* Leipzig: Xenien, 1912.

Wächtler, Eberhard, and Gisela-Ruth Engewald, eds. *Internationales Symposium zur Geschichte des Bergbaus und Hüttenwesens.* 2 vols. Freiberg: TU Bergakademie Freiberg, 1980.

Wagenbreth, Otfried. *Goethe und der Ilmenauer Bergbau.* Weimar: National Forschungs- und Gedenkstätten, 1983.

———. *Die technische Universität Bergakademie Freiberg und ihre Geschichte.* Leipzig: Deutscher Verlag für Grundstoffindustrie, 1994.

Wagenbreth, Otfried, and Eberhard Wächtler, eds. *Der Freiberger Bergbau: Technische Denkmale und Geschichte.* Leipzig: Deutscher Verlag für Grundstoffindustrie, 1985.

Wagner, Ferdinand. "Der Ober-Commissarius Graetzel zu Göttingen (1690–1770)." *Nachrichten von der Grätzel-Gesellschaft zu Göttingen* 1 (1925): 79–82.

Wagnitz, Heinrich Balthasar. *Historische Nachrichten und bemerkungen über die merkwürdigsten Zuchthäuser in Deutschland.* Halle: Gebauer, 1792.

Wakefield, Andre. "Books, Bureaus, and the Historiography of Cameralism." *European Journal of Law and Economics* 19, no. 3 (2005): 311–20.

———. "The Fiscal Logic of Enlightened German Science." In *Making Knowledge in Early Modern Europe: Practices, Objects, and Texts, 1400–1800,* edited by Pamela Smith and Benjamin Schmidt, 273–86. Chicago: University of Chicago Press, 2007.

———. "Police Chemistry." *Science in Context* 13, no. 2 (2000): 231–67.

Walker, Mack. *German Home Towns: Community, State, and Local Estate, 1648–1871.* 2nd ed. Ithaca, NY: Cornell University Press, 1998.

———. "Rights and Functions: The Social Categories of Eighteenth-Century German Jurists and Cameralists." *Journal of Modern History* 50, no. 2 (1978): 234–51.

Walter, Friedrich. *Österreichische Verfassungs- und Verwaltungsgeschichte von 1500–1955.* Vienna: Böhlau, 1972.

Walther, Rudolf. "Economic Liberalism." *Economy and Society* 13, no. 2 (1983): 178–207.

Warde, Paul. *Ecology, Economy and State Formation in Early Modern Germany.* Cambridge: Cambridge University Press, 2006.

Waszek, Norbert, ed. *Die Institutionalisierung der Nationalökonomie an deutschen Universitäten.* St. Katharinen: Scripta Mercaturae, 1988.

Weber, Max. *The Protestant Ethic and the Spirit of Capitalism.* Translated by Talcott Parsons. London: Routledge, 1930.

———. *Wirtschaft und Gesellschaft.* Tübingen: Mohr, 1922.

Weber, Wolfhard. *Innovationen im Frühindustriellen deutschen Bergbau und Hüttenwesen: Friedrich Anton von Heynitz.* Göttingen: Vandenhoeck & Ruprecht, 1976.

Webler, Heinrich. *Die Kameral-Hohe-Schule zu Lautern (1774–1784).* Speyer: n.p., 1927.

Wegert, Karl. "Contention with Civility: The State and Social Control in the German Southwest, 1760–1850." *Historical Journal* 34, no. 2 (1991): 349–69.

Wellenreuther, Hermann, ed. *Göttingen 1690–1755: Studien zur Sozialgeschichte einer Stadt.* Göttingen: Vandenhoeck & Ruprecht, 1988.

Widder, Johann Goswin. *Versuch einer vollständigen geographisch-historischen Beschreibung der Kurfürstlichen Pfalz am Rheine.* 4 vols. Frankfurt: n.p., 1786–88.

Wilckens, Christian Friedrich. *Kurtze Abhandlung von der Nothwendigkeit daß die Natur-Geschichte auf hohen Schulen gelehret werde.* Halle: Hemmerde, 1744.

Winnige, Norbert. *Krise und Aufschwung einer frühneuzeitlichen Stadt: Göttingen, 1648–1756.* Hanover: Hahnsche Buchhandlung, 1996.

Wise, M. Norton, ed. *The Values of Precision.* Princeton, NJ: Princeton University Press, 1995.

Wolin, Sheldon S. "Democracy and the Welfare State: The Political and Theoretical Connections between Staatsräson and Wohlfahrtsstaatsräson." *Political Theory* 15, no. 4 (1987): 467–500.

Wrana, Joachim, ed. *Bergakademie Freiberg: Festschrift zu ihrer Zweihundertjahrfeier am 13. November 1965.* 2 vols. Leipzig: VEB Deutscher Verlag, 1965.

Wunder, Bernd. *Geschichte der Bürokratie in Deutschland.* Frankfurt: Suhrkamp, 1986.

Zedler, Johann Heinrich, and Carl Günther Ludovici, eds. *Grosses vollständiges Universal-Lexicon aller Wissenschafften und Künste.* 64 vols. Halle: Zedler, 1732–50.

Ziekursch, Johannes. *Beiträge zur Charakteristik der preussischen Verwaltungsbeamten in Schlesien bis zum Untergang des friderizianischen Staates.* Breslau: Wohlfarth, 1907.

Zielenziger, Kurt. *Die alten deutschen Kameralisten.* Jena: Fischer, 1914.

Zimmermann, Carl F. *Obersächsische Berg-Academie.* Dresden: Hekel, 1746.

Zincke, Georg Heinrich. *Anfangsgründe der Cameralwissenschaft, worinnen dessen Grundriss weiter ausgeführt und verbessert wird.* Leipzig: Jacobi, 1755.

———. *Cameralisten-Bibliothek.* 4 vols. Leipzig: Jacobi, 1751–52.

———. *Grundriß einer Einleitung zu denen Cameral-Wissenschaften.* 2 vols. Leipzig: Fuchs, 1742–43.

Zink, Theodor. "Aus der Geschichte der pfälzischen Landwirthschaft." *Pfälzische Geschichtsblätter* 1 (1905): 3–6.

Ziolkowski, Theodore. *German Romanticism and Its Institutions.* Princeton, NJ: Princeton University Press, 1990.

INDEX